Lost Circulation

Lost Circulation
Mechanisms and Solutions

Alexandre Lavrov
SINTEF Petroleum Research, Trondheim, Norway

AMSTERDAM • BOSTON • HEIDELBERG • LONDON
NEW YORK • OXFORD • PARIS • SAN DIEGO
SAN FRANCISCO • SINGAPORE • SYDNEY • TOKYO

Gulf Professional Publishing is an imprint of Elsevier

G|P
P|Ψ

Gulf Professional Publishing is an imprint of Elsevier
50 Hampshire Street, 5th Floor, Cambridge, MA 02139, USA
The Boulevard, Langford Lane, Kidlington, Oxford, OX5 1GB, UK

Notices
Knowledge and best practice in this field are constantly changing. As new research and experience broaden our understanding, changes in research methods, professional practices, or medical treatment may become necessary.

Practitioners and researchers must always rely on their own experience and knowledge in evaluating and using any information, methods, compounds, or experiments described herein. In using such information or methods they should be mindful of their own safety and the safety of others, including parties for whom they have a professional responsibility.

To the fullest extent of the law, neither the Publisher nor the authors, contributors, or editors, assume any liability for any injury and/or damage to persons or property as a matter of products liability, negligence or otherwise, or from any use or operation of any methods, products, instructions, or ideas contained in the material herein.

British Library Cataloguing-in-Publication Data
A catalogue record for this book is available from the British Library

Library of Congress Cataloging-in-Publication Data
A catalog record for this book is available from the Library of Congress

ISBN: 978-0-12-803916-8

For information on all Gulf Professional Publishing publications
visit our website at https://www.elsevier.com/

  Working together
to grow libraries in
developing countries

ELSEVIER   Book Aid International

www.elsevier.com • www.bookaid.org

Publisher: Joe Hayton
Senior Acquisitions Editor: Katie Hammon
Senior Editorial Project Manager: Kattie Washington
Production Project Manager: Kiruthika Govindaraju
Designer: Victoria Pearson Esser

Typeset by TNQ Books and Journals

Contents

8.   Knowledge Gaps and Outstanding Issues

Preface

Lost circulation is one of the most troublesome drilling problems. Its economic cost is due to the loss of expensive drilling fluid into the formation and due to nonproductive time spent on regaining circulation. If untreated, losses may lead to well control issues, poor hole cleaning, packoffs, and stuck pipe. Choosing preventive measures at the well planning stage can mitigate lost-circulation problems significantly. If lost circulation could not be avoided through preventive measures, one has to live with losses or try to cure them. Both prevention and curing require a good understanding of what causes losses, where they occur, where and how preventive measures and cures may work, and where and why they will not work.

Despite the importance of the subject, very few books have been dedicated specifically to lost circulation since the publication of Joseph U. Messenger's *Lost Circulation* in 1981. There is a need for an up-to-date introductory text that summarizes the currently available knowledge and the remaining knowledge gaps. The present book is a step in that direction. The author's ambition was to provide a balanced overview of the current state of the art within the topic of lost circulation, taking into account its multidisciplinary nature and current industrial practices. There is currently no cure-all against lost circulation. Therefore, the book does not advocate any specific commercial product, or products of any specific vendor. There are no commercial names in this text. Hundreds of lost circulation products have been introduced on the market in the last decades. After reading this text, the reader will become better prepared for choosing effective treatments.

With the increasing share of difficult wells (deepwater, deviated, high pressure, high temperature, etc.) in the drilling portfolio, lost circulation incidents are likely to occur more often in the future. The book is intended to prepare the reader for lost circulation challenges of tomorrow. The first five chapters should equip the reader with a solid understanding of the physics and mechanics of lost circulation. The subsequent two chapters provide a broad overview of available cures and preventive techniques.

Following the introductory chapter 1, "The Challenge of Lost Circulation," chapters 2 and 3, "Stresses in Rocks" and "Natural Fractures in Rocks," introduce stresses and fractures in rocks, the two major factors that control lost circulation and are essential for understanding losses caused by natural and induced fractures.

Chapter 4, "Drilling Fluid," offers a brief introduction to drilling fluids, with special emphasis on their properties and behavior relevant to lost circulation. Understanding how a drilling fluid works is essential for understanding treatments discussed in later chapters.

Chapter 5, "Mechanisms and Diagnostics of Lost Circulation," opens with four sections on the mechanisms of lost circulation, namely losses into the porous matrix, vugs (cavities), and natural and drilling-induced fractures. Losses in different types of wells and formations are discussed (deepwater wells, deviated wells, depleted reservoirs, etc.). Some currently available theoretical models are discussed in chapter 5 too.

Mechanisms of lost circulation detailed in chapter 5 provide a background for the discussion of loss prevention techniques in chapter 6, "Preventing Lost Circulation." In particular, geological characterization provides important clues about how to mitigate the lost circulation risk by better planning the well trajectory. Chapter 6 also covers wellbore strengthening and managed pressure drilling.

Chapter 7, "Curing the Losses," covers available treatments of lost circulation once it has occurred. It explains how lost circulation materials work when they enter the thief zone. An overview of lost circulation materials and settable materials is given. Commonly accepted and, preferably, standardized testing and evaluation techniques are essential for progress in any branch of engineering, and lost circulation is no exception. Chapter 7 concludes with an overview of laboratory testing techniques currently employed in the industry to evaluate the effectiveness of different treatments and the properties of lost circulation materials.

Despite considerable progress made in the realm of lost circulation in the last 25 years, there are still many unresolved issues and outstanding challenges requiring research and development. These challenges are reviewed in chapter 8, "Knowledge Gaps and Outstanding Issues."

The preferred system of units in this book is SI. Other units are used occasionally, e.g., when quoting literature sources where those units were used. Equations are based on SI.

LIST OF ACRONYMS

APL	Annular pressure loss
BHP	Bottomhole pressure
CBP	Constant bottomhole pressure
CCS	Continuous circulation system
CWD	Casing while drilling
ECD	Equivalent circulating density
FBP	Formation breakdown pressure
FCP	Fracture closure pressure
FIP	Fracture initiation pressure
FIT	Formation integrity test
FMCD	Floating mud cap drilling
FPP	Fracture propagation pressure
FRP	Fracture reopening pressure
HPHT	High pressure, high temperature
ISIP	Instantaneous shut-in pressure
KGD	Khristianovitch−Geertsma−de Klerk
LCM	Lost circulation material
LOT	Leakoff test
LPM	Loss prevention material
MD	Measured depth
MPD	Managed pressure drilling
MWD	Measurement while drilling
NADF	Nonaqueous drilling fluids
OBM	Oil-base mud
PDC	Polycrystalline diamond compact
PKN	Perkins−Kern−Nordgren
PMCD	Pressurized mud cap drilling
PSD	Particle size distribution
RCD	Rotating control device
SBM	Synthetic-base mud
TVD	True vertical depth
UBD	Underbalanced drilling
UCS	Unconfined compressive strength
WBM	Water-base mud
WSM	Wellbore strengthening material
XLOT	Extended leakoff test

Chapter 1

The Challenge of Lost Circulation

Drilling a well is the most common way to access oil and gas resources and geothermal reservoirs. During drilling, a fluid is circulated in the well. This fluid (the *drilling fluid*) cools the drillstring, transports rock cuttings out of the well, and prevents the surrounding formation from collapse. The bottomhole pressure of the drilling fluid is kept within a certain "window." The lower bound of the wellbore pressure is usually dictated by the formation pore pressure or the minimum pressure obtained from the borehole stability analysis, whichever is greater. If the bottomhole pressure drops below the pore pressure, an influx of formation fluids into the well may occur. If the bottomhole pressure drops below the minimum value obtained from borehole stability analysis, the formation may cave in.

The upper operational bound of the bottomhole pressure is chosen so as to avoid lost circulation. *Lost circulation* is a situation where less fluid is returned from the wellbore than is pumped into it. When lost circulation occurs, some drilling fluid is lost into the formation. Lost circulation gives rise to nonproductive time spent on regaining circulation. According to Ref. [1], lost circulation was responsible for more than 10% of nonproductive time spent when drilling in the Gulf of Mexico between 1993 and 2003. The inability to cure losses and resume drilling may, in the worst case, necessitate sidetracking or abandoning the well.

The economic impact of lost circulation includes, in addition, the costs of the lost drilling fluid and of the treatment used to cure the problem. According to one estimate, the cost of drilling fluids amounts to 25%−40% of total drilling costs [2]. Given that both regular drilling fluids and lost circulation materials are often quite expensive, the direct economic impact of losing these substances into the formation may be substantial. The cost issue is especially relevant for oil-based muds that are usually more costly than water-based fluids.

In addition to the direct economic impact (cost of expensive drilling fluid and nonproductive time), lost circulation may cause additional drilling problems. In particular, the reduced rate of returns may impair cuttings transport out of the well. This leads to poor hole cleaning, especially in deviated and horizontal wells [3]. Poor hole cleaning may eventually result in pack-offs and stuck pipe.

Lost Circulation. http://dx.doi.org/10.1016/B978-0-12-803916-8.00001-7

Losing drilling fluid into the formation in the pay zone increases formation damage as pores and fractures in the reservoir rock become plugged with particles present in the drilling mud (barite, bentonite, cuttings, solids used as lost circulation material, etc.). The formation damage created by lost circulation needs to be removed before production can start, which leads to additional costs.

Severe cases of lost circulation may lead to well control problems. In particular, the mud column disappearing into the formation may reduce the fluid pressure in the well, which will enable the influx of formation fluids, in particular gas, into the well. This may eventually lead to a kick or borehole collapse. Lost circulation in tophole sections may lead to shallow water flow events.

Given the scope of its negative consequences, lost circulation has been identified as "one of the drilling industry's most singular problems" [4]. According to some estimates, the annual cost of lost circulation problems, including the cost of materials and the rig time, is around one billion dollars globally [5,6].

Lost circulation in the overburden can be equally as bad as in the reservoir, even though formation damage is of no concern there. If losses are not treated properly and drilling proceeds without first sealing the thief zone, subsequent cement jobs can be compromised. The quality of well cementing depends crucially on placing the cement column all the way up to the target height. If an unplugged thief zone exists against the annulus to be cemented, cement slurry may escape into this zone during the cement job, and the cemented length of the annulus will be shorter than planned. Remedial cementing can be employed to cure the problem, but this will increase nonproductive time and incur extra costs.

Lost circulation is common in geothermal drilling (Boxes 1.1 and 1.2) [7,8]. Large fracture apertures (on the order of cm) often cause severe or total losses while drilling the overburden or the reservoir. According to Ref. [9], lost circulation problems are responsible for 10% of well costs in mature geothermal fields and often more than 20% of well costs in exploration wells in the United States. In Iceland, an analysis revealed that lost circulation or hole collapse was the primary cause of drilling troubles in 18 out of 24 wells in the Hengill Geothermal Area [10]. These problems may further lead to cement losses into the formation during subsequent well cementing.

Wells drilled in fields with elevated geothermal gradient are often prone to losses caused by cooling. When the relatively cold drilling fluid coming from the surface contacts the much hotter formation, the rock contracts and the hoop stress around the hole becomes smaller—ie, less compressive. The rock is then easier to fracture because of this effect. Ballooning and losses observed in some Gulf of Mexico wells are attributed to this effect [11].

Lost circulation is common in naturally fractured formations. Severe or total losses are common in carbonate rocks in the Middle East [12].

BOX 1.1 Lost Circulation in a Geothermal Well in Iceland

Lost circulation is a common problem in geothermal drilling, where it is exacerbated by high temperatures and hard rocks. A sequence of lost circulation events while drilling a geothermal well in the Krafla field in Iceland in 2008–09 was described by Pálsson et al. in their paper "Drilling of the well IDDP-1" (*Geothermics*, 2014, 49, 23–30).

The well, IDDP-1, was part of the Iceland Deep Drilling Project and was originally designed to reach a 4500-m depth. Severe problems were encountered during drilling, and the well had to be sidetracked every time it approached magma at 2100 m. Mud losses were experienced often, becoming worse as the depth increased. Also, the well became increasingly more unstable and washed out with depth, which affected the hole cleaning and reduced the rate of penetration that already was low due to the hard rock.

Minor mud losses occurred in the first 1000 m while drilling for the intermediate 18-5/8″ casing. The losses were cured with lost circulation material.

Losses of 20 L/s were then experienced at 1432 m. The problem was cured by cementing the thief zone.

Losses in excess of 60 L/s were experienced at 2043 m. The losses could not be cured. The weighted drilling fluid was replaced with water, and drilling continued.

At 2101 m, the bottomhole assembly broke. After unsuccessful fishing, a cement plug was placed, and the well was sidetracked.

Upon the sidetrack, total losses occurred at 2054 m. A cement plug was placed in the well, and the drilling continued from 2060 m. Losses resumed at 2067 m, and total loss of circulation occurred at 2076 m.

Continued problems with stuck pipe and unsuccessful fishing attempts at 2103 m forced a second sidetrack. Mud losses continued after the sidetrack, and total loss of circulation occurred at 2071 m. Drilling was terminated upon reaching magma at 2100 m. The well was then tested and completed.

In a naturally fractured carbonate field in Iran, mud losses were reported in 35% of drilled wells [3]. In Saudi Arabia, 32% of wells in the naturally fractured carbonate Khuff Formation experience ballooning, while 10% experience lost circulation [13].

The best way to deal with lost circulation is to prevent it from happening altogether in the first place. In practice, however, this may be difficult to achieve. Nevertheless, technological improvements in formation characterization and drilling fluid design enable the prevention of losses in many wells. Preventing lost circulation requires that the mechanics and physics of this drilling problem are fully understood.

The most obvious way to prevent lost circulation is to keep the downhole pressure sufficiently low—ie, below the upper operational pressure bound. In practice, however, it is not quite obvious how this upper bound should be chosen. In competent intact formations, the upper bound of the operational

BOX 1.2 Losses in Geothermal Well WK204 in New Zealand

A dramatic sequence of lost circulation events, culminating in a blowout, occurred during drilling of an investigation well at Wairakei Geothermal Field in New Zealand in 1960. The case history was described by Bolton et al. in their paper "Dramatic incidents during drilling at Wairakei Geothermal Field, New Zealand" (*Geothermics*, 2009, 38, 40−47).

The well was drilled near a fault. The initial design was as follows:

- 406-mm casing to 27 m (surface casing);
- 298-mm casing to 122 m (anchor casing);
- 219-mm casing to 305 m or deeper, if the conditions allow.

Major losses were experienced while drilling for the 298-mm casing. Cementing the casing required six times the annular volume, indicating that cement went into thief zones.

While drilling for the 219-mm casing, a thief zone was encountered at 134 m. The zone was sealed, and drilling continued to the target depth of 305 m with full returns. The decision was made to continue drilling deeper than 305 m.

Major losses started at 350 m. Attempts to cure the losses with increasingly coarser lost circulation materials were unsuccessful. The drill bit dropped by 1.5 m at 373 m. Further developments eventually resulted in stopping the pumps.

In the aftermath of the events, the lost circulation experienced at 350−373 m was attributed to the well penetrating a high-pressure high-temperature (HPHT) zone near the fault. Buildup of temperature and pressure in the hole after the pumps were stopped eventually led to breakdown of the seal set at 134 m, and a blowout.

bottomhole pressure is often set equal to the minimum in situ stress (minus some safety margin). The upper pressure bound is often called *fracture pressure* (We shall prefer the term "fracturing pressure" rather than "fracture pressure" in this book to avoid possible confusion with the fluid pressure inside a fracture. Fracture gradient, routinely used in drilling practice, is the fracturing pressure divided by the height of the mud column (psi/ft or kPa/m). The fracture gradient, in general, increases with depth since the bulk density of rocks, in general, increases with depth. Deviations from this trend, however, are possible. In particular, depleted formations may exhibit significantly lower pore pressure gradient and fracture gradient.) in this case, since an induced fracture will not propagate if the wellbore pressure stays below the minimum in situ stress. However, as we shall see, mud can be lost not only into induced fractures, but also into high-permeability zones (gravel, unconsolidated sand, etc.), large cavities, and natural fractures. The minimum in situ stress plays only a minor role in those scenarios. Also, induced fractures do not always cause lost circulation. As long as the induced fracture is short and narrow, losses might be acceptable or not noticeable at all.

Therefore, it would be more appropriate to call the upper operational pressure bound "lost-circulation pressure" rather than "fracturing pressure." Lost-circulation pressure means simply the bottomhole pressure above which lost circulation will occur, without reference to any specific (and often unknown) mechanism.

The lost-circulation pressure is a major uncertainty, even in competent rocks. This uncertainty is increased in depleted or complex reservoirs where pore pressure and stress distributions are rarely known. In naturally fractured rocks, the lost-circulation pressure depends on both the orientation and the aperture of natural fractures. Apertures of natural fractures may indeed be so small that the drilling fluid will not be able to enter them. Different fracture orientations mean that different fractures will open and cause losses at different wellbore pressures. Since there is usually a great variety in both apertures and orientations of natural fractures, it makes sense to consider a *spectrum* of lost-circulation pressures, rather than a single value, in such formations. This shift of paradigm may help in situations where the upper pressure bound estimated from formation integrity and leakoff tests performed on a short openhole section below the casing shoe is later found to be misguiding. Indeed, a short open hole pressurized in such tests provides only a sample of the natural fractures that may be encountered by the drill bit during subsequent drilling. The results of the tests are therefore not always representative of what lies ahead.

The profiles of pore pressure and fracturing pressure (or, alternatively, pore pressure gradient and fracture gradient) versus depth determine the maximum length of the interval that can be drilled with the same mud weight. Thus, they determine the location of casing points along the well. This is illustrated for a fictitious vertical onshore well in Fig. 1.1 by plotting the upper and lower operational pressure profiles. The static bottomhole pressure is shown with the *dotted line*. Inclined parts of the *dotted line* must pass through zero pressure at the surface since the annular pressure is zero at the surface (This is true for conventional drilling where the circulation system is opened to the atmosphere. In managed pressure drilling, a backpressure may be applied.). Jumps in the *dotted line* signify changes in the mud weight. The largest possible lengths of the intervals are shown by asterisks. Indeed, moving the casing point located at D_1 deeper along the hole would bring the static bottomhole pressure below the pore pressure in the lower part of the interval. Increasing the mud weight to remedy this problem would increase the slope of the *dotted line*, which would violate the upper pressure bound in the upper part of the interval. This example shows that the lost-circulation pressure and the pore pressure (or the borehole stability limit) play crucial roles in setting up the casing program.

An alternative representation of the problem is possible in terms of the *pore pressure gradient* and *lost-circulation pressure gradient*. This is illustrated for

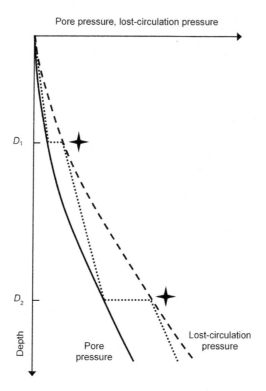

FIGURE 1.1 Casing points determined from the lower (pore pressure or borehole stability limit) and upper (lost-circulation pressure) operational pressure bounds along the well. The static bottomhole pressure is shown with a *dotted line*. *Asterisks* denote casing points.

a fictitious vertical onshore well in Fig. 1.2. The result—ie, the locations of the casing points—is, of course, the same as in Fig. 1.1.

The ability to predict the lost-circulation pressure is therefore crucial for optimizing the casing program. Figs. 1.1 and 1.2 also suggest that if the lost-circulation pressure could be increased, longer intervals could be drilled with the same mud weight. It would reduce the number of casing points and increase the well diameter at the target depth. This is the motivation for applying special treatments to increase the lost-circulation pressure in the open hole ("wellbore strengthening"), discussed in chapter "Preventing Lost Circulation."

Even if the bottomhole pressure stays below the lost-circulation pressure during normal drilling, pressure surges during trips and connections may exceed this upper bound and cause lost circulation. When a connection is made, circulation is suspended. During connection, the drilling fluid starts developing gel strength. When circulation is resumed after connection, the pressure needs to be increased sufficiently to break the gel. This may lead to

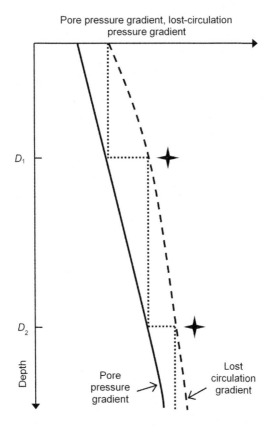

FIGURE 1.2 Casing points determined from the gradient profiles along the well. The static mud weight is shown with a *dotted line. Asterisks* denote casing points.

substantial pressure changes during connections. If the operational pressure window is narrow—eg, in a deepwater well—the reduction of pressure before connection may lead to a formation fluid influx, and the pressure surge after connection may lead to lost circulation.

Trips may cause formation fluid influx when pulling out of the hole, and lost circulation when running in hole. Preventing influxes and lost circulation during trips can be achieved, for example, by optimizing the drilling fluid rheology.

Drilling through depleted formations is often required in order to access deeper reservoirs. Depleted formations are prone to mud losses. In some wells drilled in depleted formations, losses on the order of thousands of barrels have been reported [14]. The minimum horizontal stress is usually reduced in depleted reservoirs (chapter: Stresses in Rocks). This reduction affects the operational pressure window by reducing the fracturing pressure and thereby increasing the risk of mud losses.

Deviated or horizontal wells are prone to lost circulation. The operational pressure window is narrow in such wells. In some cases, the window may close altogether as the inclination increases. In extended reach wells, the problem is additionally aggravated by increasing annular pressure losses along the horizontal section. Since the fracturing pressure remains approximately the same at the same depth, the bottomhole pressure will eventually exceed it.

A considerable share of mud losses occur when running casing or pumping cement. Running the casing pipe generates an excessive bottomhole pressure that can lead to formation breakdown. During cementing, high density and rheology of cement result in an elevated bottomhole pressure, likely to be the highest pressure the formation is ever exposed to during well construction. This may lead to lost circulation during a cement job.

Another example of formation where losses are common is subsalt rubble zones [15]. Such formations are often represented by relatively weak and/or fractured shale. It has been argued that fractures in these shales are caused by deformation in the adjacent salt throughout geological history [15]. Pore pressure in the subsalt shale can be either lower or higher than the pore pressure in the salt. The former scenario is the case, for example, in the Hassi Messaoud field, where severe losses were experienced [16]; the latter scenario is the case, for example, in the Gulf of Mexico, with pore pressure vs. depth schematically shown in Fig. 1.3. High pore pressure in shale below the salt is

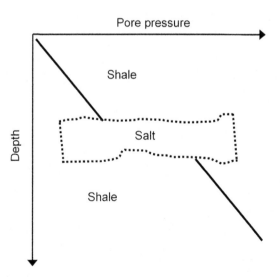

FIGURE 1.3   Pore pressure profile (*solid line*) above and below a salt body typical of some Gulf of Mexico wells. The salt body is shown with a *dotted line*. Overpressure in the subsalt shale is caused by the salt "trapping" the pore pressure. *Based on Sweatman R, Faul R, Ballew C. New solutions for subsalt-well lost circulation and optimized primary cementing. SPE paper 56499 presented at the 1999 SPE annual technical conference and exhibition held in Houston, Texas; 3–6 October 1999.*

caused by the salt "trapping" the pore pressure. It results in a narrow margin between the pore pressure and the lost-circulation pressure in shale. As with other fractured rocks, filling fractures in the subsalt shale with cement or lost circulation material is not an easy task. According to one report from 1999, "the cost of drilling formations approximately 1500 ft above the salt to approximately 1500 ft below the salt have reached several million dollars" [15]. The nonproductive time associated with such intervals can be weeks. As an example, more than half of the wells drilled in the Hassi Messaoud field experienced total losses in subsalt shale [16].

Lost circulation is exacerbated in deepwater drilling. The fracture gradient is often quite low in deepwater wells [17] (Fig. 1.4). This results in a narrow operational pressure window in such wells. Exceeding the fracture gradient can lead to mud losses.

Lost circulation is a multidisciplinary challenge. Combatting lost circulation requires a complex approach that includes rock mechanical analysis, careful well trajectory planning, optimization of drilling fluid rheology and composition, optimization of loss prevention and lost circulation materials, and optimization of drilling hydraulics [18].

Considerable advances in petroleum-related rock mechanics, including hydraulic fracture mechanics, over the past decades have improved our understanding of lost circulation. Better methods of mud loss prediction and more effective treatments make it possible to drill wells that would be impossible to drill a few decades ago. At the same time, increasingly more difficult drilling conditions are encountered as ever deeper reserves are involved in exploration and production. This results in persistence of the lost-circulation challenge to this day.

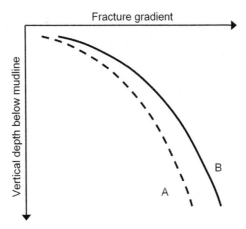

FIGURE 1.4 Fracture gradient as a function of vertical depth in deepwater drilling: water depth in case A is greater than in case B.

Lost-circulation incidents happen regularly all over the world. In the future, the incidence of lost circulation is likely to increase. Five types of challenging wells are becoming increasingly common in oil and gas industry and will continue to be so in future:

- deepwater wells;
- wells in depleted reservoirs;
- deviated, horizontal, and extended-reach wells;
- HPHT wells;
- combinations of the above.

As we will see in subsequent chapters, the risk of lost circulation is increased in all five of these types of wells.

The aim of this book is to prepare the reader for the lost-circulation challenges of tomorrow. This is done by providing an up-to-date explanation of lost-circulation mechanisms and of current industrial practices. We start our journey into the realm of lost circulation by reviewing some basic concepts of rock mechanics in the next chapter.

REFERENCES

[1] Rehm B, Schubert J, Haghshenas A, Paknejad AS, Hughes J, editors. Managed pressure drilling. Houston (TX): Gulf Publishing Company; 2008.

[2] Lécolier E, Herzhaft B, Néau L, Quillien B, Kieffer J. Development of a nanocomposite gel for lost circulation treatment. SPE paper 94686 presented at the SPE European formation damage conference held in Scheveningen, The Netherlands; 25–27 May 2005.

[3] Abdollahi J, Carlsen IM, Mjaaland S, Skalle P, Rafiei A, Zarei S. Underbalanced drilling as a tool for optimized drilling and completion contingency in fractured carbonate reservoirs. SPE/IADC paper 91579 presented at the 2004 SPE/IADC underbalanced technology conference and exhibition held in Houston, Texas, USA; 11–12 October 2004.

[4] Sanders MW, Scorsone JT, Friedheim JE. High-fluid-loss, high-strength lost circulation treatments. SPE paper 135472 presented at the SPE deepwater drilling and completions conference held in Galveston, Texas, USA; 5–6 October 2010.

[5] Al Maskary S, Abdul Halim A, Al Menhali S. Curing losses while drilling & cementing. SPE paper 171910 presented at the Abu Dhabi international petroleum exhibition and conference held in Abu Dhabi, UAE; 10–13 November 2014.

[6] Droger N, Eliseeva K, Todd L, Ellis C, Salih O, Silko N, et al. Degradable fiber pill for lost circulation in fractured reservoir sections. IADC/SPE paper 168024 presented at the 2014 IADC/SPE drilling conference and exhibition held in Fort Worth, Texas, USA; 4–6 March 2014.

[7] Pálsson B, Hólmgeirsson S, Guðmundsson Á, Boasson HÁ, Ingason K, Sverrisson H, et al. Drilling of the well IDDP-1. Geothermics 2014;49:23–30.

[8] Bolton RS, Hunt TM, King TR, Thompson GEK. Dramatic incidents during drilling at wairakei geothermal field, New Zealand. Geothermics 2009;38:40–7.

[9] Finger J, Blankenship D. Handbook of best practices for geothermal drilling. Sandia National Laboratories; 2010. Contract No.: SAND2010–6048.

[10] Sveinbjornsson BM, Thorhallsson S. Drilling performance, injectivity and profuctivity of geothermal wells. Geothermics 2014;50:76−84.

[11] Tare UA, Whitfill DL, Mody FK. Drilling fluid losses and gains: case histories and practical solutions. SPE paper 71368 presented at the 2001 SPE annual technical conference and exhibition held in New Orleans, Louisiana; 30 September−3 October 2001.

[12] Savari S, Whitfill DL. Managing losses in naturally fractured formations: sometimes nano is too small. SPE/IADC paper 173062 presented at the SPE/IADC drilling conference and exhibition held in London, United Kingdom; 17−19 March 2015.

[13] Ameen MS. Fracture and in-situ stress patterns and impact on performance in the Khuff structural prospects, eastern offshore Saudi Arabia. Mar Petrol Geol 2014;50:166−84.

[14] Willson SM, Edwards S, Heppard PD, Li X, Coltrin G, Chester DK, et al. Wellbore stability challenges in the deep water, Gulf of Mexico: case history examples from the Pompano field. SPE paper 84266 presented at the SPE annual technical conference and exhibition held in Denver, Colorado, USA; 5−8 October, 2003.

[15] Sweatman R, Faul R, Ballew C. New solutions for subsalt-well lost circulation and optimized primary cementing. SPE paper 56499 presented at the 1999 SPE annual technical conference and exhibition held in Houston, Texas; 3−6 October, 1999.

[16] Ferras M, Galal M, Power D. Lost circulation solutions for severe sub-salt thief zones. Paper AADE-02-DFWM-HO-30 presented at the AADE 2002 technology conference "Drilling & completion fluids and waste management," held at the Radisson Astrodome, Houston, Texas, April 2−3, 2002 in Houston, Texas; 2002.

[17] Power D, Ivan CD, Brooks SW. The top 10 lost circulation concerns in deepwater drilling. SPE paper 81133 presented at the SPE Latin American and Caribbean petroleum engineering conference held in Port-of-Spain, Trinidad, West Indies; 27−30 April 2003.

[18] Whitfill DL, Hemphill T. All lost-circulation materials and systems are not created equal. SPE paper 84319 presented at the SPE annual technical conference and exhibition held in Denver, Colorado, USA; 5−8 October 2003.

Chapter 2

Stresses in Rocks

The stress state in the reservoir and around the borehole, and rock failure caused by mechanical or thermal stresses, play central roles in lost circulation. These concepts are reviewed in this chapter.

2.1 STRESS, STRENGTH, AND FAILURE

When mechanical loads are applied to the rock, or when the rock is heated or cooled, stresses are induced. *Stress* is defined as the force acting on a surface of unit area. If we cut out an imaginary cube inside the rock with the faces normal to the coordinate axes, the stress on each face can be decomposed into the normal stress and the shear stress. The latter can be decomposed once again into two components acting along two coordinate directions parallel to the face as shown in Fig. 2.1. Normal stresses are denoted by σ, and shear stresses are denoted by τ, in Fig. 2.1. The first index in the shear stress—eg, "x" in

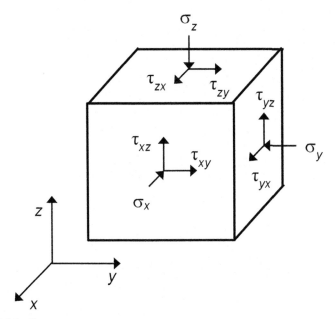

FIGURE 2.1 Stress components.

Lost Circulation. http://dx.doi.org/10.1016/B978-0-12-803916-8.00002-9

13

τ_{xy}—denotes the face on which the stress is acting. The second index denotes the direction of the stress. The directions of normal stresses in Fig. 2.1 correspond to compressive stresses. Normal stresses induce longitudinal deformations. Compressive normal stresses induce contraction, whereas tensile normal stresses induce extension. A sign convention is needed for normal stresses. In rock mechanics, compressive stresses are usually assumed positive, and tensile stresses negative.

Equilibrium conditions require that the shear stresses with the same but permuted indexes be equal, for instance $\tau_{xy} = \tau_{yx}$, and similarly for other shear stress components.

The nine stress components make up a second-order *stress tensor*:

$$\sigma = \begin{pmatrix} \sigma_x & \tau_{xy} & \tau_{xz} \\ \tau_{yx} & \sigma_y & \tau_{yz} \\ \tau_{zx} & \tau_{zy} & \sigma_z \end{pmatrix} \qquad [2.1]$$

Due to the symmetry properties of the shear stresses, the stress tensor is symmetric and thus has six independent components.

Rotating the cube in Fig. 2.1 would, in general, change the values of normal and shear stresses (except in the case of a uniform, so-called *hydrostatic* compression or tension in which case there are no shear stresses, and all normal stresses have the same magnitude). There exists one special orientation of the cube such that there are no shear stresses on its faces (Fig. 2.2). Normal stresses acting on the cube faces in this case are called *principal stresses*. They are denoted by σ_1, σ_2, and σ_3, with σ_1 being the greatest of the three (the most compressive one) and σ_3 being the smallest of the three (the tensile or least compressive one). The directions of these stresses are called *principal axes*.

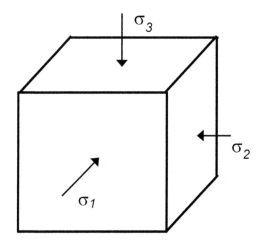

FIGURE 2.2 Principal stresses.

In the coordinate system coincident with the principal axes, the stress tensor has only diagonal entries:

$$\boldsymbol{\sigma} = \begin{pmatrix} \sigma_1 & 0 & 0 \\ 0 & \sigma_2 & 0 \\ 0 & 0 & \sigma_3 \end{pmatrix} \qquad [2.2]$$

In fluid-saturated porous media, stresses are carried partially by the rock grains and partially by the saturating fluid. The part of the stress carried by the rock grains is called the *effective stress*. Since typical saturating fluids, such as oil or gas, cannot carry shear stresses, the concept of the effective stress only makes sense for normal stresses. An effective normal stress, σ', is related to the total normal stress, σ, and the pore pressure, P_p, as follows:

$$\sigma' = \sigma - \alpha P_p \qquad [2.3]$$

where α is a dimensionless coefficient called the *effective stress coefficient* (or the *Biot effective stress coefficient*). In soils and unconsolidated or weak rocks, the value of α is close to 1. For other types of rocks, such as sandstone or shale, the value of α is typically smaller than 1.

The effective stresses are responsible for the deformation of the porous media. For porous media, the constitutive equations of linear poroelasticity are given by

$$E_d \varepsilon_x = \sigma'_x - v_d \left(\sigma'_y + \sigma'_z \right)$$

$$E_d \varepsilon_y = \sigma'_y - v_d \left(\sigma'_z + \sigma'_x \right)$$

$$E_d \varepsilon_z = \sigma'_y - v_d \left(\sigma'_x + \sigma'_y \right)$$

$$2G\gamma_{xy} = \tau_{xy} \qquad [2.4]$$

$$2G\gamma_{yz} = \tau_{yz}$$

$$2G\gamma_{zx} = \tau_{zx}$$

where $\varepsilon_x, \varepsilon_y, \varepsilon_z$ are normal strains in the x, y, and z directions, respectively; $\gamma_{xy}, \gamma_{yz}, \gamma_{zx}$ are shear strains; E_d and v_d are the Young's modulus and the Poisson's ratio in the drained regime, respectively (also known as the Young's modulus and the Poisson's ratio of the rock framework); and G is the shear modulus. Eq. [2.4] are an analogue of Hooke's law of classical linear elasticity.

In addition to controlling the deformation of porous media, the effective stresses also enter the failure criteria. For a given rock, the value of α used in the failure criteria can be different from that used in poroelasticity equations.

A rock can only tolerate stresses up to a certain limit. When that limit is exceeded, the rock fails. Rock failure can be either in tension or in shear.

Tensile failure happens when the effective tensile stress acting on some plane exceeds the *tensile strength*. Since the lowest (ie, the most tensile) principal stress is σ_3, the tensile failure criterion is given by:

$$\sigma'_3 = -T_0 \qquad [2.5]$$

where T_0 is the tensile strength of the rock. The effective stress coefficient, α, used to evaluate the effective stress in Eq. [2.5] is often set equal to 1.0 for both unconsolidated and consolidated rocks [1].

The tensile strength can be determined in laboratory by a direct tension test. In this test, a cylindrical specimen of the rock is subject to uniaxial tension along its axis. Dividing the maximum load that the specimen can sustain by the area of the cross section normal to its axis yields T_0. In practice, the direct tension test is difficult to perform. Instead, an indirect tension test is used. One of the most popular indirect tension tests is the so-called *Brazilian test*, in which a disk or cylinder of rock is subjected to compression along one of the diameters shown as a dashed line in Fig. 2.3. In the central area of the disk, tensile stress is produced in the direction normal to the loading diameter. When this stress exceeds the tensile strength, the disk splits along the loading diameter.

Apart from tensile failure, a rock can fail by shear failure. This happens when the shear stress exceeds a certain limit on certain favorably oriented planes (failure planes). This limit, $\tau_{\max}$, depends on the effective normal stress σ'_n acting on the failure plane, the concept known as Mohr's hypothesis [1]:

$$|\tau_{\max}| = f\left(\sigma'_n\right) \qquad [2.6]$$

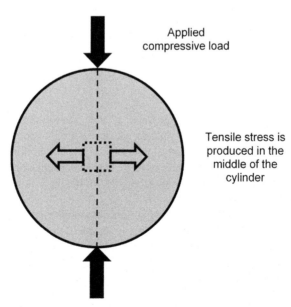

FIGURE 2.3 Schematic plot of Brazilian test. The *dashed line* indicates the loading diameter.

In the $\sigma'-\tau$ coordinates, the Mohr's hypothesis can be represented using the failure envelope shown as the solid line in Fig. 2.4. The stress states located between the failure envelope and the σ'-axis are admissible. The stress states located above or to the left of the failure envelope cannot be attained without breaking the rock. The rock fails when a point in the stress space reaches the failure envelope during loading. It is convenient to illustrate this by means of a *Mohr's circle* which is shown as a dashed line in Fig. 2.4. The Mohr's circle in Fig. 2.4 is supported by the minimum and maximum effective stresses and shows the stress state on planes with different orientations. While the failure envelope describes the properties of the rock itself, the Mohr's circle describes the actual state of stress induced in the rock by loading. In the situation shown in Fig. 2.4, the Mohr's circle does not cross the failure envelope. Thus, no failure occurs (note that the minimum principal stress, σ'_3, is compressive in Fig. 2.4, and thus no tensile failure occurs at that point, either). If, as a result of loading, the Mohr's circle is expanded by increasing the maximum effective stress or decreasing the minimum effective stress, the Mohr's circle may eventually touch the failure envelope, resulting in shear failure. Shear failure can be induced also by moving the Mohr's circle as a whole leftward by increasing the pore pressure and thus reducing the effective stresses. Eventually, the circle will touch the envelope in this case, again resulting in failure.

In practical rock mechanics analysis, the curved failure envelope of Fig. 2.4 is often approximated with a straight line, as shown in Fig. 2.5. The resulting failure criterion is known as the *Mohr—Coulomb failure criterion*. The intercept of the failure envelope with the τ-axis is known as the *cohesion*, S_0. The slope of the failure envelope is characterized by the *coefficient of internal friction*, μ. Thus, the Mohr—Coulomb criterion has three parameters: S_0, μ, and T_0.

If a cylindrical rock specimen is subjected to compressive loading along the cylinder's axis while its side surface remains stress-free, failure occurs

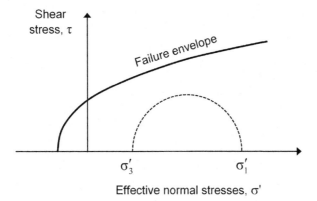

FIGURE 2.4 Failure envelope and Mohr's circle.

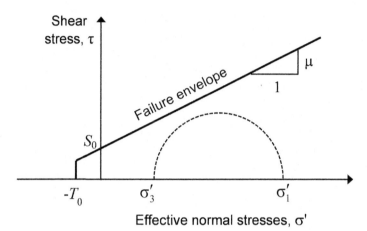

FIGURE 2.5   Mohr–Coulomb failure criterion.

when the compressive stress reaches the *unconfined compressive strength* (*UCS*). UCS is an important rock mechanical property of the rock and is linked to the cohesion and the coefficient of internal friction as follows:

$$\text{UCS} = 2S_0 \tan\left(\frac{\pi}{4} + \frac{\varphi}{2}\right) \tag{2.7}$$

where $\tan\varphi = \mu$ (the angle φ is known as the *angle of internal friction*). In order to obtain both parameters S_0 and μ, a uniaxial compressive test is not enough, since it only provides one point on the Mohr–Coulomb failure envelope. At least one triaxial test is needed to obtain a second point on the failure envelope in order to calculate both S_0 and μ. A triaxial test with $\sigma_1 > \sigma_2 = \sigma_3 > 0$ is usually used to this end.

Fig. 2.4 illustrates an important aspect of the Mohr's theory: failure is not affected by the intermediate principal stress, σ_2. This assumption is not strictly correct, and more elaborate failure criteria have been proposed to account for the effect of the intermediate principal stress; for example, see Ref. [1] for more details.

2.2 IN SITU STRESSES

Rock stress is one of the in situ factors controlling lost circulation. In situ stresses can change as a result of production from or injection into the reservoir. Production- and injection-induced changes of in situ stresses will be discussed in Sections 2.4 and 2.5. In this section, we focus on the original in situ state of stress.

To define the in situ stresses means to define all three principal stresses, with their magnitudes and directions. Since deformation and failure of rocks depend on the effective rather than the total stresses, the in situ pore pressure plays a crucial role here.

In an ideal case of sediments deposited as horizontal layers, and horizontal Earth's surface, the total vertical stress at a given depth, D, is one of the principal stresses and can be estimated from the weight of the overlaying rocks as follows:

$$\sigma_v = \int_0^D \rho_b(z)g\,dz \qquad [2.8]$$

where ρ_b is the bulk density of the rock at depth z; g is the acceleration of gravity. Assuming full saturation, the bulk density is given by: $\rho_b = \phi\rho_f + (1 - \phi)\rho_s$, where ϕ, ρ_f, and ρ_s are the rock porosity, the pore fluid density, and the density of the mineral grains, respectively [2].

Assessing horizontal in situ stresses is considerably more difficult, even in this simplest case of horizontal layers. In the absence of tectonic forces, the horizontal stresses are the result of lateral confinement only. In this case, and assuming the rock behaves linear elastically, the horizontal in situ stresses can be evaluated by setting $\sigma_z' = \sigma_v', \sigma_x' = \sigma_y' = \sigma_H' = \sigma_h', \varepsilon_x = \varepsilon_y = 0$ in Eq. [2.4], which yields

$$\sigma_h' = \sigma_H' = \frac{v_d}{1 - v_d}\sigma_v' \qquad [2.9]$$

with the maximum, σ_H', and minimum, σ_h', horizontal effective stresses being equal in this case. In Eq. [2.9], v_d is the drained Poisson's ratio of the rock. The effective vertical stress is given by

$$\sigma_v' = \sigma_v - \alpha P_p \qquad [2.10]$$

where P_p is the pore pressure and σ_v is given by Eq. [2.8].

Eq. [2.9] shows that the vertical stress is the largest compressive stress in this simple model since $0 < v_d < 0.5$. This stress regime is known as *extensional* $(\sigma_v > \sigma_H)$. If faults are present, this stress regime would promote *normal faulting*.

Tectonic forces may produce an in situ stress state with the largest stress being horizontal. If the vertical stress is the smallest principal stress, the stress regime is called *compressional* $(\sigma_H \geq \sigma_h > \sigma_v)$. It results in *reverse faulting* (thrust faulting). If both the largest and the smallest in situ stresses are horizontal $(\sigma_H > \sigma_v > \sigma_h)$, the stress-regime is *strike-slip faulting*.

In addition to the basic tectonic regimes—ie, normal, reverse, and strike-slip—combinations are possible, including normal faulting with some strike-slip component, strike-slip faulting with some normal or reverse component, or reverse faulting with some strike-slip component.

Apart from tectonic forces, elevated horizontal stresses can be caused by geological history, in particular if the formation underwent uplift caused by erosion of the overburden. The horizontal stresses acting before the uplift remain locked in the rock unless they are gradually dissipated by stress relaxation caused by viscoelastic behavior of the rock.

Therefore, in reality, the coefficient in front of σ'_v in Eq. [2.9] is not necessarily equal to $v_d/(1 - v_d)$, and the relationship between the horizontal and the vertical effective stress could more accurately be written as follows:

$$\sigma'_h = \sigma'_H = K'\sigma'_v \qquad [2.11]$$

The coefficient K' typically varies between 1 and 10 at shallow depth, and between 0.2 and 1.5 at great depth [1].

Crude estimates of horizontal stresses can be obtained by using empirical relations, such as the Breckels—van Eekelen equations for the total minimum horizontal stress at the US Gulf Coast [1]:

$$\sigma_h = 0.0053D^{1.145} + 0.46(P_p - P_{pn}) \text{ at } D < 3500 \text{ m}$$
$$\sigma_h = 0.0264D - 31.7 + 0.46(P_p - P_{pn}) \text{ at } D > 3500 \text{ m} \qquad [2.12]$$

where D is the depth; P_p is the pore pressure; and P_{pn} is the "normal pore pressure" corresponding to the pore pressure gradient of 10.5 MPa/km.

Contrary to the simplified model given by Eq. [2.11], the horizontal stresses are often not equal: $\sigma_H \neq \sigma_h$.

So far, we have assumed that one of the in situ principal stresses is vertical. Even though this is true in many cases, the directions of principal stresses may be gravely affected by surface topography. In particular, near a slope, the vertical direction might not coincide with any of the principal stress direction (eg, see Ref. [3]). In the valley, the horizontal stress can be larger than the vertical stress even in the absence of tectonic forces.

Fig. 2.6 illustrates the impact of structural features on the in situ stresses onshore. In deepwater drilling, horizontal stress anomalies similar to those

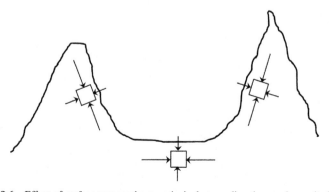

FIGURE 2.6   Effect of surface topography on principal stress directions and magnitudes.

shown in Fig. 2.6 are observed near surfaces of escarpments, as opposed to intrabasin areas. Not taking the reduced horizontal stresses into account when designing a well may result in lost circulation incidents when drilling the top-hole section [4]. Directions of in situ stresses in the reservoir can be affected by nearby faults, domes, and other geological features.

To illustrate complexities of in situ stress states, let us consider stress distributions around salt bodies. Drilling through salt is often accompanied with borehole instabilities and on exiting the salt, lost circulation. A numerical geomechanical analysis performed in Ref. [5] provides a valuable insight into the problems experienced while drilling salt. Salt cannot sustain shear stresses over long periods. Stress relaxation leads to gradual reduction of shear stresses in salt, so that the state of stress gradually approaches hydrostatic compression, with $\sigma_v \approx \sigma_H \approx \sigma_h$. The inability to sustain shear stresses for long periods makes salt similar to a very viscous fluid which, as any fluid, would eventually approach hydrostatic equilibrium. The in situ stresses in surrounding rocks (eg, shale) may on the other hand be far from hydrostatic. In the Gulf of Mexico, the tectonic regime is normal faulting, and the far-field stresses in rocks around the salt are such that $\sigma_v > \sigma_H \approx \sigma_h$. The discrepancy between the stress fields inside the salt body and outside it leads to significant complexity of the stress field around the boundary of the salt. Around a salt body that is close to spherical in shape, the stresses are altered compared with their far-field values, as shown in Fig. 2.7. Horizontal stresses are reduced above and below the salt, which explains the reduced fracturing pressure experienced on exiting the salt while drilling. In the sideburden—ie, in the rocks laterally adjacent to the salt—the stress changes are more

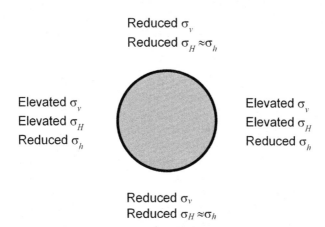

Reduced σ_v
Reduced $\sigma_H \approx \sigma_h$

Elevated σ_v Elevated σ_v
Elevated σ_H Elevated σ_H
Reduced σ_h Reduced σ_h

Reduced σ_v
Reduced $\sigma_H \approx \sigma_h$

FIGURE 2.7 Alteration of in situ stresses around an idealized, spheroidal salt body (shown in gray). *Based on the results of Fredrich JT, Coblentz D, Fossum AF, Thorne BJ. Stress perturbations adjacent to salt bodies in the deepwater Gulf of Mexico. SPE paper 84554 presented at the SPE Annual Technical Conference and Exhibition held in Denver, Colorado, USA, October 5–8, 2003.*

complex: horizontal stresses in those areas become anisotropic, even if the far-field horizontal stresses are isotropic. In addition, principal stresses may be rotated in those areas, with the vertical direction not being a principal direction any more [5]. These stress changes have a detrimental effect on borehole stability.

Reduced vertical stresses above salt bodies indicate that using density profiles to evaluate the vertical in situ stress, Eq. [2.8], would overestimate the in situ vertical stress in these areas. The reduction in the vertical stress above the salt under natural conditions is an example of the *arching effect*, similar to the one observed during depletion (Section 2.4). Note, however, that the arching effect in Fig. 2.7 is caused solely by natural stress redistribution due to stress relaxation before any reservoir depletion. In particular, since salt is unable to sustain shear stresses, the vertical stress needs to be redistributed so that some part of the vertical load that would otherwise be borne by the salt is carried by the surrounding rock.

The spherical geometry of the salt body in Fig. 2.7 is a strong idealization. Numerical analyses with more realistic geometries confirm that reduced horizontal in situ stresses persist right above and right below the salt [6]. This is consistent with lost circulation incidents experienced when entering or exiting the salt. A reduction of the fracture gradient by 3 ppg (0.36 g/cm^3) underneath the salt was reported for a well in the Gulf of Mexico [6].

If the shape of the salt body is not spheroidal (eg, in the case of a salt sheet), stress perturbations around the salt are less severe. According to the numerical results of Fredrich et al. [5], the hydrostatic stress magnitude in a spheroidal salt body lies between the far-field values of σ_v and $\sigma_H \approx \sigma_h$. In a salt body with lateral extension larger than its height by a factor of 4 or more, the hydrostatic stress in the salt is close to the far-field vertical stress.

In addition to reduced horizontal stresses, the rock below the salt is often damaged or broken. These *rubble zones* further contribute to mud losses on exiting the salt.

The uncertainty about the values of σ_H, σ_h, and their directions call for experimental methods that would enable measurement of these parameters in the field. Some of the methods currently used in the industry will be discussed in Section 2.7.

Numerical modeling is another way to obtain estimates of both vertical and horizontal stresses in and around the reservoir. The use of numerical modeling requires that the rocks (reservoir, overburden, and sideburden) are properly characterized in terms of rock mechanical properties, or at least the values of these parameters are constrained to within sufficiently narrow margins. Due to scarcity of core material, particularly from the caprock, such constraining is often difficult. In spite of this difficulty, examples of numerical evaluation of rock stresses are available. For instance, finite-element stress analysis was used to evaluate stresses in and around a salt diapir in the deepwater Gulf of Mexico [6].

2.3 PORE PRESSURE

Along with in situ stresses, pore pressure is another factor important for lost circulation. In a normally pressured formation, the pore pressure is hydrostatic—ie, it is given by $\rho_f g D_{TVD}$, where ρ_f is the fluid density, g is the acceleration of gravity, and D_{TVD} is the true vertical depth. In practice, deviations from the hydrostatic pore pressure profile are common. Pore pressure below the hydrostatic is usually called "subnormal," while pore pressure above the hydrostatic is called "abnormal." Subnormal pore pressure can be the result of the geological history, or it can be caused by reservoir depletion. Abnormal pore pressure can be caused, for example, by rapid compaction of a low-permeability rock (typically shale). Subnormally pressured formations represent a lost circulation risk. Abnormally pressure formation may cause fluid influx into the well if the bottomhole pressure becomes lower than the in situ pore pressure during drilling.

2.4 DEPLETION-INDUCED STRESS CHANGES

Lost circulation is exacerbated in depleted reservoirs. Understanding why this is so requires understanding of depletion-induced stress changes.

Eq. [2.3] suggests that a change in the pore pressure (eg, its reduction caused by oil production) will increase the value of the effective stress if the total stress remains unchanged. An injection (during water flooding, for example) will do the opposite—ie, reduce the effective stress. Increasing or decreasing effective stresses can close or open natural fractures available in the reservoir. In addition, decreasing effective stresses can promote slip on faults (and fractures), since the critical shear stress needed to induce slip depends on the effective normal stress acting on the fault plane. According to the Coulomb criterion, the critical shear stress for slip initiation is given by [7]:

$$\tau_c = \mu(\sigma_n - P_p) \qquad [2.13]$$

where μ is the coefficient of friction; σ_n is the total normal stress on the fault plane; and P_p is the pore pressure. Note that Eq. [2.13] implies that the Biot effective stress coefficient, α, is equal to 1, an assumption commonly made in failure criteria.

In addition to effective stresses, depletion also affects total stresses in the reservoir and to a lesser extent in the caprock. In order to see why total stresses are affected by depletion, consider a very simple model of a horizontal reservoir of constant height and infinite extent in horizontal directions. Assume that both the reservoir and the caprock are linear elastic, pressure changes uniformly in the entire reservoir during depletion, and there are no pressure changes in the caprock. In this case, as the pore pressure changes by $\Delta P_p < 0$, the total vertical stress in the reservoir will remain unchanged and equal to the weight of the overburden. The effective vertical stress in the reservoir will thus increase by $\Delta\sigma_v' = -\alpha\Delta P_p > 0$ due to depletion, according to Eq. [2.3]. Since the reservoir is constrained in the lateral (horizontal) directions, the effective horizontal stresses will increase by $\Delta\sigma_H' = \Delta\sigma_h' = -\alpha\nu_d\Delta P_p/(1 - \nu_d) > 0$, according to Eq. [2.9]. Thus, again using Eq. [2.3], the *total* horizontal stress will change by $\Delta\sigma_H = \Delta\sigma_h = \alpha(1 - 2\nu_d)\Delta P_p/(1 - \nu_d)$. The latter expression is negative, since $\Delta P_p < 0$ and $0 < \nu_d < 0.5$. Hence, depletion-induced changes of the total and effective stresses in the reservoir depend on the mechanical properties of the rock, in particular the Biot effective stress coefficient. *The total horizontal stress in the reservoir becomes less compressive during depletion.*

The above argument suggests that the effective horizontal stresses increase during depletion. This will result in closing of vertical fractures and thus might reduce mud losses into preexisting fractures when drilling through such a depleted reservoir with small overbalance. On the other hand, a decrease in the total horizontal stresses means that it will be easier to fracture the formation if the overbalance becomes sufficiently high. This may exacerbate the problem of mud losses caused by induced fractures. This is often observed while drilling in depleted fields.

The increase in the vertical effective stress in the reservoir will promote closing of horizontal fractures.

The total vertical stress remaining constant and the total horizontal stress being reduced during depletion suggest that normal faulting will be promoted in the reservoir if the overall stress environment is extensional. This indeed has been observed in the North Sea [7]. Slip on critically oriented normal faults and shear fractures may increase their permeability if the fractures open up as asperities on their faces slide over each other (the phenomenon known as *dilatancy*). This may happen if the normal stress on the shear plane is sufficiently low. If the normal stress is high, asperities will be crushed, and the permeability will decrease during the slip. Shear-induced fracture permeability changes are discussed in detail in chapter "Natural Fractures in Rocks." Here, it is sufficient to point out that if dilatancy increases the permeability of a reactivated fault or a shear fracture, it may enhance mud losses into such fractures if they are intersected during drilling. Activation of normal faults caused by depletion has been proposed as an explanation for continuously high productivity of chalk reservoirs in the North Sea maintained despite their significant compaction [7].

The simple model discussed above assumes that the reservoir has infinite horizontal dimensions. In reality, reservoirs have limited lateral extent. A reservoir may also have a quite complex shape and structure. Coupled numerical geomechanical simulations are required in order to evaluate stress dynamics caused by depletion of real reservoirs. However, some general conclusions can still be drawn by using simplified models of such reservoirs. In particular, an ellipsoidal reservoir with two major axes pointing in horizontal directions, and the minor axis pointing in the vertical direction, has often been used as such a simple geomechanical model (eg, Refs [1,8]).

In general, depletion-induced stress dynamics in the reservoir can be described by using nondimensional stress path coefficients given by [1]:

$$\gamma_v^{(res)} = \frac{\Delta\sigma_v^{(res)}}{\Delta P_p^{(res)}} \qquad [2.14]$$

$$\gamma_H^{(res)} = \frac{\Delta\sigma_H^{(res)}}{\Delta P_p^{(res)}} \qquad [2.15]$$

$$\gamma_h^{(res)} = \frac{\Delta\sigma_h^{(res)}}{\Delta P_p^{(res)}} \qquad [2.16]$$

where the superscript, (res), indicates that the changes of stresses and pore pressure within the reservoir are considered. $\gamma_v^{(res)}, \gamma_H^{(res)}, \gamma_h^{(res)}$ describe how a change of the reservoir pressure by 1 MPa affects, respectively, the total vertical, total maximum horizontal, and total minimum horizontal stresses in the reservoir. In the case of an elastic reservoir, infinitely large in the horizontal directions, $\gamma_v^{(res)} = 0$ and $\gamma_H^{(res)} = \gamma_h^{(res)} = \alpha(1 - 2v_d)/(1 - v_d)$. In reality, finite horizontal dimensions of the reservoir result in $\gamma_v^{(res)}$ being nonzero and positive. This is the result of the *arching effect*. The coefficient $\gamma_v^{(res)}$ is known as the arching coefficient. Due to arching, the total vertical stress is somewhat reduced in the reservoir during depletion; ie, the reservoir becomes partially shielded from the overburden weight. The vertical stress is redistributed onto the sideburden.

The relation $\gamma_H^{(res)} \approx \gamma_h^{(res)} > \gamma_v^{(res)} > 0$ signifies again that normal faulting is promoted in the reservoir during depletion. The values of $\gamma_v^{(res)}, \gamma_H^{(res)}, \gamma_h^{(res)}$ vary for different reservoirs. For instance, $\gamma_h^{(res)} = 0.7$ has been reported for the crest of the Valhall chalk reservoir in the North Sea [7]. In general, $\gamma_v^{(res)}, \gamma_H^{(res)}, \gamma_h^{(res)}$ depend on the reservoir geometry and are affected by the stiffness contrast between the reservoir and the surrounding rocks. In particular, the arching effect is promoted, ie, $\gamma_v^{(res)}$ is larger, if the reservoir is softer than the surrounding rocks [1].

In addition to the stress changes in the reservoir considered so far, depletion may also alter the stresses in the overburden, sideburden, and underburden. Stress path coefficients can be defined in the overburden, sideburden, and underburden, similar to the reservoir stress path coefficients—eg, for the overburden:

$$\gamma_v^{(ovb)} = \frac{\Delta\sigma_v^{(ovb)}}{\Delta P_p^{(res)}} \qquad [2.17]$$

$$\gamma_H^{(ovb)} = \frac{\Delta\sigma_H^{(ovb)}}{\Delta P_p^{(res)}} \qquad [2.18]$$

$$\gamma_h^{(ovb)} = \frac{\Delta\sigma_h^{(ovb)}}{\Delta P_p^{(res)}} \qquad [2.19]$$

where the superscript, (ovb), indicates that the changes of stresses within the overburden are considered. Contrary to the reservoir, the values of $\gamma_H^{(ovb)}, \gamma_h^{(ovb)}$ are negative in the overburden. It means that the total horizontal stresses in the overburden somewhat increase during depletion. The reason for this increase is that the reservoir tries to contract in the lateral direction under depletion, and this induces contraction in the overburden in the lateral direction, resulting in extra compressive horizontal stresses.

Depletion-induced stress changes in the overburden are smaller than in the reservoir: $\gamma_v^{(ovb)} < \gamma_v^{(res)}$, $\left|\gamma_h^{(ovb)}\right| < \gamma_h^{(res)}$, at least as long as the pore pressure decrease occurs predominantly in the reservoir. For instance, estimates for Valhall indicate that $\left|\gamma_h^{(ovb)}\right|$ is only c. 1% of $\gamma_h^{(res)}$ [7]. The values of $\gamma_v^{(ovb)}, \gamma_H^{(ovb)}, \gamma_h^{(ovb)}$ depend on the ratio between the reservoir height and its lateral extent, and approach zero when the reservoir becomes very large in the horizontal directions.

The total vertical stress in the overburden can decrease during depletion due to the effect of the Earth's surface and because the total vertical stress should be continuous across the reservoir−overburden boundary. If the pore pressure is unchanged in the overburden during depletion, the effective vertical stress will somewhat decrease as well, by the same amount as the total vertical stress does. These changes in the vertical stresses might facilitate opening of horizontal fracture in the overburden, and might promote induced horizontal fractures during drilling if the overbalance is sufficiently high.

Since the vertical stresses (total and effective) decrease in the overburden while the horizontal stresses (total and effective) increase, reverse faulting is promoted in the overburden if the tectonic environment is compressional. If the tectonic environment is extensional, normal faulting in the overburden is suppressed during depletion.

Due to poroelastic effects caused by increased horizontal stresses, the pore pressure may slightly increase in the overburden during reservoir depletion if the permeability of the overburden is sufficiently small. (In such case, the caprock is in an approximately undrained hydraulic condition [8].) This will further complicate the geomechanical picture.

In the sideburden, the total vertical stress will tend to increase due to the arching effect. Assuming that the sideburden is considerably less permeable than the reservoir, the pore pressure in the sideburden is not affected by the reservoir depletion. Therefore, the effective vertical stress in the sideburden will increase, too. The total horizontal stress normal to the sideburden—reservoir boundary will tend to decrease during reservoir depletion, as it does in the reservoir itself, since this stress must be continuous across the reservoir—sideburden boundary. Since, by our assumption, the pore pressure does not change in the sideburden, this means that the effective horizontal stress will decrease in the sideburden, unlike in the reservoir. These stress changes will promote opening of existing vertical fractures and facilitate induced fractures during drilling in the sideburden of a depleted reservoir. In addition, in an extensional tectonic regime, normal faulting will be promoted in the sideburden since the total vertical stress increases and the total horizontal stress decreases there. In a compressional regime, depletion-induced stresses will suppress reverse faulting and thus stabilize reverse faults in the sideburden.

A summary of possible stress changes in the reservoir, overburden, and sideburden during reservoir depletion is given in Table 2.1. When constructing Table 2.1, it was assumed that the contrast of elastic properties between the reservoir and the surrounding rocks is sufficiently small, the reservoir is approximately horizontal, and the pore pressure reduction is limited to the reservoir. These stress dynamics are further illustrated in Fig. 2.8.

TABLE 2.1 Stress Changes During Depletion

Location	$\Delta\sigma_v$	$\Delta\sigma'_v$	$\Delta\sigma_H, \Delta\sigma_h$	$\Delta\sigma'_H, \Delta\sigma'_h$
Reservoir	−	+	−	+
Overburden	−	−	+	+
Sideburden	+	+	−	−

A plus designates an increase, with the stress becoming more compressive, while a minus designates a decrease, with the stress becoming less compressive.

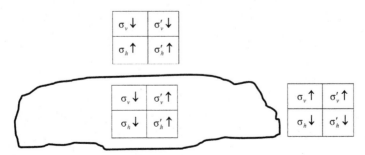

FIGURE 2.8 Depletion-induced stress changes in and around a reservoir. The *solid line* shows the boundary of the reservoir. *Arrow pointing up* means the stress becomes more compressive. *Arrow pointing down* means the stress becomes less compressive.

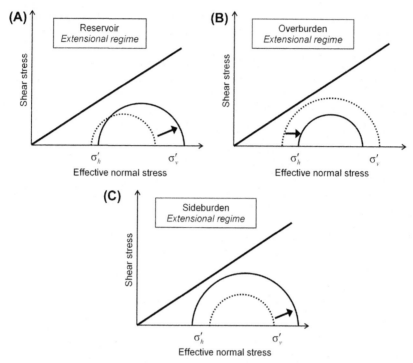

FIGURE 2.9 Schematic illustration of effective stress alterations in reservoir (A), overburden (B), and sideburden (C) during depletion. Extensional (normal faulting) in situ stress regime is assumed. Pore pressure is assumed to remain constant outside the reservoir. Mechanical properties of the reservoir and surrounding rocks are assumed to be the same. *Dotted Mohr circle*: before depletion. *Solid Mohr circle*: after depletion. *Arrows* indicate change of the Mohr circle due to depletion. Subscripts v and h refer to the vertical and minimum horizontal stresses, respectively.

The stress dynamics shown in Table 2.1 and Fig. 2.8 bring about changes in the Mohr circles shown in Figs. 2.9–2.11.

Depletion-induced stress changes discussed above may induce fracturing, faulting, or both in the reservoir or in the surrounding formations, or may make fractures and faults more prone to reactivation during drilling. Some of the possible fracturing and faulting scenarios caused by depletion are shown in Table 2.2.

In addition to the scenarios in Table 2.2 discussed above, depletion affects the total horizontal stresses: the total horizontal stresses in the reservoir and the sideburden decrease, while those in the overburden slightly increase. These stress changes lead to a decrease in the fracture initiation and fracture reopening stresses in the depleted reservoir and in the sideburden, while in the overburden, the fracture initiation and fracture reopening pressures slightly increase. As a result, depletion increases the risk of lost circulation due to induced fracturing when drilling through the reservoir and the sideburden.

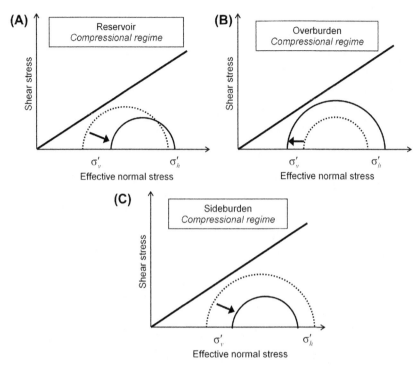

FIGURE 2.10 Schematic illustration of effective stress alterations in reservoir (A), overburden (B), and sideburden (C) during depletion. Compressional (reverse faulting) in situ stress regime is assumed. Pore pressure is assumed to remain constant outside the reservoir. Mechanical properties of the reservoir and surrounding rocks are assumed to be the same. *Dotted Mohr circle*: before depletion. *Solid Mohr circle*: after depletion. Solid arrow indicates change of the Mohr circle due to depletion. Subscripts v and h refer to the vertical and minimum horizontal stresses, respectively.

The issue of lost circulation in depleted reservoirs is often exacerbated by the presence of undepleted low-permeability shale layers within the reservoir. The pore pressure maintained in such shales can be greater than the minimum in situ stress in the drained reservoir rock [1]. Thus, staying below the minimum in situ stress in the depleted reservoir rock may require a mud weight that is not sufficient to ensure borehole stability in the shale. Vice versa, maintaining a mud weight high enough to avoid borehole instabilities in shale may fracture the sandstone. Wellbore strengthening and other preventive measures improve safety and economics of drilling in depleted reservoirs (see chapter "Preventing Lost Circulation").

As mentioned above, the effect of changing reservoir pressure on the stress state in the caprock is typically smaller than on the stress state in the reservoir itself [9]. Therefore, fracture and fault reactivation scenarios in the overburden shown in Table 2.2 will be able to fully develop only after the onset of fault (or fracture) reactivation in the reservoir.

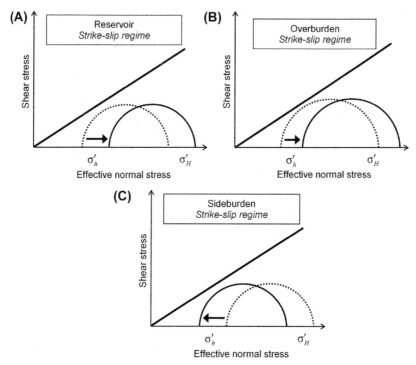

FIGURE 2.11   Schematic illustration of effective stress alterations in reservoir (A), overburden (B), and sideburden (C) during depletion. Strike-slip in situ stress regime is assumed. Pore pressure is assumed to remain constant outside the reservoir. Mechanical properties of the reservoir and surrounding rocks are assumed to be the same. *Dotted Mohr circle*: before depletion. *Solid Mohr circle*: after depletion. *Solid arrows* indicate change of the Mohr circle due to depletion. Subscripts v and h refer to the vertical and minimum horizontal stresses, respectively.

2.5 INJECTION-INDUCED STRESS CHANGES

Injection of a fluid into a depleted reservoir (eg, water during water flooding) increases the pore pressure. It is hardly possible, however, to bring the reservoir precisely to the pressure state it was in prior to the start of production. But even if it were theoretically possible to exactly rebuild the pore pressure distribution that existed before the onset of depletion, the stress state would not return to its original predepletion state. Observations suggest that the stress path coefficients can be smaller during injection than they were during depletion. In particular, while $\gamma_h^{(res)} = \Delta\sigma_h^{(res)}/\Delta P_p^{(res)}$ is often about 0.5–0.8 during depletion [1,7,10], it can reportedly be almost zero during subsequent injection into the depleted reservoir [11]. The following is usually used to explain the difference between stress paths during depletion and subsequent injection: From a geomechanical viewpoint, depletion corresponds

TABLE 2.2 Examples of Possible Effects of Reservoir Depletion on Fractures and Faults Under Different Tectonic Stress Regimes. "Pore pressure" refers to the reservoir pore pressure; pore pressure in the surrounding rocks is assumed to remain unchanged

Stress Regime	Reservoir	Overburden	Sideburden
Extensional (normal faulting)	Slip reactivation on normal faults; closing of vertical and horizontal fractures	Stabilization of normal faults; possible slip propagation along normal faults from reservoir into overburden; opening of horizontal fractures; closing of vertical fractures	Slip reactivation on normal faults; opening of vertical fractures; closing of horizontal fractures
Compressional (reverse faulting)	Stabilization of reverse faults; closing of vertical and horizontal fractures	Slip reactivation on reverse faults; opening of horizontal fractures; closing of vertical fractures	Stabilization of reverse faults; opening of vertical fractures; closing of horizontal fractures
Strike-slip	Closing of vertical and horizontal fractures	Opening of horizontal fractures; closing of vertical fractures	Opening of vertical fractures; closing of horizontal fractures

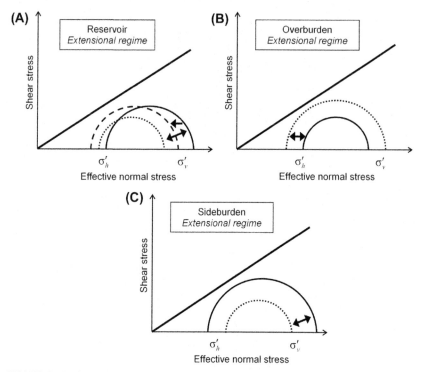

FIGURE 2.12 Schematic illustration of effective stress alterations in the reservoir (A), over-burden (B), and sideburden (C) during depletion and subsequent injection into a reservoir. Extensional (normal faulting) in situ stress regime is assumed. Pore pressure is assumed to remain constant outside the reservoir. Mechanical properties of the reservoir and surrounding rocks are assumed to be the same. *Dotted Mohr circle*: before depletion and after injection, assuming un-changed stress path (reversible deformation). *Solid Mohr circle*: after depletion. *Dashed Mohr circle*: after injection, assuming zero stress path. Subscripts *v* and *h* refer to the vertical and minimum horizontal stresses, respectively. *Double arrow* indicates depletion and injection with the same stress path. *Single arrow* indicates injection with reduced (zero) stress path.

to reservoir loading (increase of effective stresses). Subsequent injection into a depleted reservoir corresponds to unloading. Zero (or low) stress path co-efficients are believed to be due to plastic deformation created in the rock during depletion.

Assume, as before, that during both depletion and subsequent injection, the pore pressure only changes inside the reservoir, the reservoir is horizontal, and the stiffness of the reservoir is not much different from the over-, under-, and side-burden. Under these assumptions, stress changes under depletion and injection are illustrated in Figs. 2.12–2.14 for extensional (normal faulting), compressional (reverse faulting), and strike-slip regimes, respectively. Note that *irreversible* deformation is assumed to occur only in the reservoir in Figs. 2.12–2.14, and thus the stress path in the overburden and sideburden is the same during injection as it is

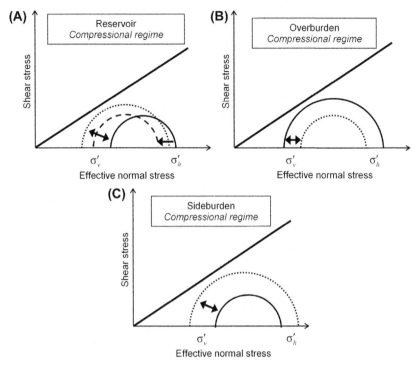

FIGURE 2.13 Schematic illustration of effective stress alterations in the reservoir (A), overburden (B), and sideburden (C) during depletion and subsequent injection into a reservoir. Compressional (reverse faulting) in situ stress regime is assumed. Pore pressure is assumed to remain constant outside the reservoir. Mechanical properties of the reservoir and surrounding rocks are assumed to be the same. *Dotted Mohr circle*: before depletion and after injection, assuming unchanged stress path (reversible deformation). *Solid Mohr circle*: after depletion. *Dashed Mohr circle*: after injection, assuming zero stress path. Subscripts *v* and *h* refer to the vertical and minimum horizontal stresses, respectively. *Double arrow* indicates depletion and injection with the same stress path. *Single arrow* indicates injection with reduced (zero) stress path.

during depletion. Two cases are illustrated for the reservoir in panel A in each of these figures. Stress changes with zero reservoir stress path coefficients during injection are shown by single arrows. Stress changes with unchanged original stress path coefficients during injection are shown by double arrows. The latter case corresponds to reversible reservoir deformation during depletion—injection, by which the Mohr circle returns to its original position (dotted). The stress paths in the overburden (caprock) and sideburden are assumed to be the same under depletion and injection; ie, deformation is reversible in those rocks. In Fig. 2.14, the stress paths for the minimum and maximum horizontal stresses are assumed to be equal, $\gamma_H^{(res)} = \gamma_h^{(res)}$.

The reservoir parts of Figs. 2.12 and 2.14 can be found in [12] for a specific field case. Figs. 2.12A and 2.14A demonstrate that nonzero stress path

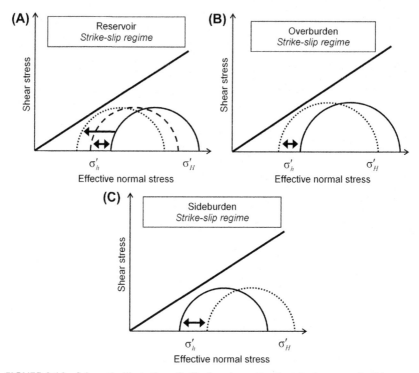

FIGURE 2.14 Schematic illustration of effective stress alterations in the reservoir (A), over-burden (B), and sideburden (C) during depletion and subsequent injection into a reservoir. Strike-slip in situ stress regime is assumed. Pore pressure is assumed to remain constant outside the reservoir. Mechanical properties of the reservoir and surrounding rocks are assumed to be the same. *Dotted Mohr circle*: before depletion and after injection, assuming unchanged stress path (reversible deformation). *Solid Mohr circle*: after depletion. *Dashed Mohr circle*: after injection, assuming zero stress path. Subscripts *v* and *h* refer to the vertical and minimum horizontal stresses, respectively. *Double arrow* indicates depletion and injection with the same stress path. *Single arrow* indicates injection with reduced (zero) stress path.

coefficients reduce the risk of fault reactivation in the reservoir in normal and strike-slip regimes, as pointed out in [9,12].

Irreversibility of rock deformation exemplified by zero (or very low) stress path coefficients would have profound effects on fractures and faults in the reservoir and in the caprock. In particular, if normal faults had been reactivated in the reservoir during depletion, they will not be able to return to their initial state during injection because it is not possible to recreate the predepletion state of stress by simply repressurizing the reservoir to the original pressure.

Injection-induced stress changes discussed above may induce fracturing or faulting in the reservoir and the surrounding formations, or they may make fractures and faults more prone to reactivation during drilling. Some possible effects of injection on fractures and faults are summarized in Table 2.3.

TABLE 2.3 Examples of Possible Effects of Injection Into a Depleted Reservoir Under Different Tectonic Stress Regimes. "Pore pressure" refers to the reservoir pore pressure; pore pressure in the surrounding rocks is assumed to remain unchanged

Stress Regime	Reservoir	Overburden	Sideburden
Extensional (normal faulting)	Possible slip on normal faults if the reservoir stress path is sufficiently low during injection; opening of vertical and horizontal fractures		Closing of vertical fractures
Compressional (reverse faulting)			Closing of vertical fractures
Strike-slip	Opening of vertical fractures if the reservoir stress path is sufficiently low during injection		Closing of vertical fractures

If the stress path during injection is smaller than during depletion, the total stresses in the reservoir cannot be restored to their predepletion level by increasing the reservoir pressure. Thus, the increased risk of mud losses caused by induced vertical fractures may persist while drilling a reservoir where pressure has been restored by fluid injection.

Apart from the stress changes caused by poroelastic effects during pressure buildup under injection, water flooding may have an additional impact on stresses due to the cooling of the formation. The injected water can be considerably colder than the reservoir rock. This will induce thermal stresses in the reservoir, namely, the total stresses will be somewhat reduced (in addition to their possible increase by poroelastic effects if the stress path is different from zero). A crude estimate of the reduction in total stresses in the reservoir caused by cooling is given by $\Delta\sigma^{(\text{thermal})} = K_d\beta\Delta T < 0$, neglecting possible thermal-induced pore pressure changes [8]. Here, $K_d = E_d/[3(1 - 2v_d)]$ is the bulk modulus of the rock framework (the drained bulk modulus); ΔT is the temperature change in the reservoir ($\Delta T < 0$); β is the linear coefficient of thermal expansion of the rock. The latter is on the order of 10^{-5} 1/K for sedimentary rocks. The drained bulk modulus can vary substantially and may be on the order of $0.1-10$ GPa for reservoir sandstones [1]. For a sandstone with $K_d = 10$ GPa, a decrease of the reservoir temperature by $10°C$ will produce a reduction in the reservoir total stresses on the order of 1 MPa. Reduced total horizontal stresses will increase the risk of mud losses. Therefore, the cooling-induced reduction of σ_H, σ_h will lower the upper bound of the drilling window.

2.6 STRESSES IN THE NEAR-WELL AREA

When a borehole is drilled, the rock surrounding it is no longer supported by the rock that originally occupied the borehole. This leads to stress redistribution in the near-well area. Drilling fluid provides some support to the rock around the borehole. However, unless the in situ stress state in the plane normal to the borehole axis is perfectly isotropic ($\sigma_H = \sigma_h$ in the case of a vertical well), it is not possible to restore the stresses around the borehole to exactly the same state that existed before drilling. The borehole thereby acts as a stress concentrator in the formation. The rock around the borehole is prone to failure if the stress concentration becomes too high. The stress state in the near-well area plays a crucial role in lost circulation and in the design of treatments and preventive measures. This section provides a brief summary of near-well stresses and borehole failure modes. A more in-depth treatment of failure specific for lost circulation is given in chapter "Mechanisms and Diagnostics of Lost Circulation."

The stress distribution around a borehole is affected by in situ stress anisotropy, borehole orientation, and pore pressure distribution. The simplest case is the one of a vertical borehole, isotropic horizontal in situ stresses ($\sigma_H = \sigma_h$) and constant pore pressure. Constant pore pressure, independent of the distance from the borehole and of the azimuth, may exist if an impermeable filter cake is rapidly deposited on the borehole wall and protects the pore fluid in the rock from hydraulic communication with the well ("impermeable borehole wall"). If the rock is linear elastic, the total stresses are then given by (eg, see Ref. [1])

$$\sigma_r = \left(1 - \frac{r_w^2}{r^2}\right)\sigma_h + \frac{r_w^2}{r^2}P_w \qquad [2.20]$$

$$\sigma_\theta = \left(1 + \frac{r_w^2}{r^2}\right)\sigma_h - \frac{r_w^2}{r^2}P_w \qquad [2.21]$$

$$\sigma_z = \sigma_v \qquad [2.22]$$

where r_w is the radius of the wellbore and P_w is the fluid pressure in the well. The stresses in Eqs. [2.20]–[2.22] are written in a cylindrical coordinate system associated with the borehole. The z-axis is along the borehole axis. The azimuthal coordinate θ specifies the angle in the borehole cross section normal to its axis; r is the radial coordinate—ie, the radial distance from the borehole axis. This is illustrated in Fig. 2.15 where the directions of the radial stress, σ_r, and the circumferential stress (hoop stress), σ_θ, are indicated by arrows. Eqs. [2.20]–[2.22] are obtained under the assumption of plane strain conditions. The assumption of plane strain is valid as long as the borehole is sufficiently long.

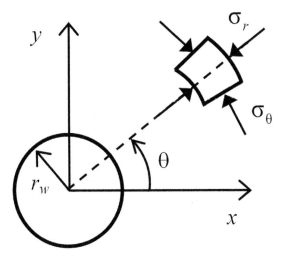

FIGURE 2.15 Directions of near-well stresses in a cylindrical coordinate system.

At the borehole wall ($r = r_w$), the total radial stress is equal to the borehole pressure:

$$\sigma_r = P_w \qquad [2.23]$$

The hoop stress is given by

$$\sigma_\theta = 2\sigma_h - P_w \qquad [2.24]$$

at the borehole wall, which signifies the stress concentration mentioned above.

The hoop stress is maximum at the borehole wall, and decreases with r according to Eq. [2.21]. This is a direct consequence of linear elastic behavior of the rock assumed when deriving Eqs. [2.20]–[2.22]. It suggests that the rock is most likely to fail at the borehole wall. More sophisticated models of rock behavior—eg, nonlinear elasticity or plasticity—result in more complex (and more realistic) stress distributions with the maximum σ_θ no longer being at the borehole wall but being shifted into the formation.

If the borehole wall is permeable, and the pore pressure varies only with r, analytical solutions for near-well stresses can be obtained—see Ref. [1], for example, for derivation and final results. In this case, the stresses at the borehole wall ($r = r_w$) are given by

$$\sigma_r = P_w \qquad [2.25]$$

$$\sigma_\theta = 2\sigma_h - P_w + \alpha \frac{1 - 2v_d}{1 - v_d}[P_p(r_w) - P_{p0}] \qquad [2.26]$$

$$\sigma_z = \sigma_v + \alpha \frac{1 - 2v_d}{1 - v_d}[P_p(r_w) - P_{p0}] \qquad [2.27]$$

where P_{p0} is the initial pore pressure before drilling (the far-field pore pressure); $P_p(r_w)$ is the actual, currently existing pore pressure at the borehole wall.

If there is no filter cake; ie, perfect hydraulic communication exists between the well and the pore space, $P_p(r_w) = P_w$. Note that the coefficient, $\alpha(1 - 2v_d)/(1 - v_d)$, appearing in Eqs. [2.26] and [2.27] is the same as the reservoir stress path coefficient in a linear elastic model of a depleting reservoir introduced in Section 2.4.

The case of isotropic in situ stresses acting normal to the borehole axis considered so far in this section illustrates the idea of stress concentration quite clearly. The isotropic case is, however, an idealization since, in general, all three in situ stresses may have different magnitudes, and the borehole can be drilled in a direction different from the principal stress directions. The elastic solution in such generic case is quite cumbersome and can be found, for example, in Ref. [1].

In a less generic case of all in situ stresses being different $\sigma_v \neq \sigma_H \neq \sigma_h$, pore pressure being constant and independent of r and θ (impermeable borehole wall), and the borehole being vertical (and thus parallel to the in situ principal stress σ_v), the solution is given by [1]:

$$\sigma_r = \frac{\sigma_H + \sigma_h}{2}\left(1 - \frac{r_w^2}{r^2}\right) + \frac{\sigma_H - \sigma_h}{2}\left(1 + \frac{3r_w^4}{r^4} - \frac{4r_w^2}{r^2}\right)\cos 2\theta + P_w\frac{r_w^2}{r^2} \quad [2.28]$$

$$\sigma_\theta = \frac{\sigma_H + \sigma_h}{2}\left(1 + \frac{r_w^2}{r^2}\right) - \frac{\sigma_H - \sigma_h}{2}\left(1 + \frac{3r_w^4}{r^4}\right)\cos 2\theta - P_w\frac{r_w^2}{r^2} \quad [2.29]$$

$$\sigma_z = \sigma_v - 2v_d(\sigma_H - \sigma_h)\frac{r_w^2}{r^2}\cos 2\theta \quad [2.30]$$

$$\tau_{r\theta} = -\frac{\sigma_H - \sigma_h}{2}\left(1 - \frac{3r_w^4}{r^4} + \frac{2r_w^2}{r^2}\right)\sin 2\theta \quad [2.31]$$

$$\tau_{rz} = \tau_{\theta z} = 0 \quad [2.32]$$

where the polar angle, θ, is measured from the direction of σ_H. At the borehole wall ($r = r_w$), the stresses are then given by:

$$\sigma_r = P_w \quad [2.33]$$

$$\sigma_\theta = \sigma_H + \sigma_h - 2(\sigma_H - \sigma_h)\cos 2\theta - P_w \quad [2.34]$$

$$\sigma_z = \sigma_v - 2v_d(\sigma_H - \sigma_h)\cos 2\theta \quad [2.35]$$

$$\tau_{r\theta} = \tau_{rz} = \tau_{\theta z} = 0 \quad [2.36]$$

Unlike the case of isotropic stresses acting normal to the borehole axis, the hoop stress, σ_θ, in Eq. [2.34] is a function of azimuth, θ.

The maximum values of the hoop stress are obtained at the two opposite points located at $\theta = \pi/2$ and $\theta = 3\pi/2$. The diameter joining these two points is parallel to the minimum in situ stress normal to the borehole axis, σ_h. The maximum value of the hoop stress is given by:

$$\sigma_{\theta max} = 3\sigma_H - \sigma_h - P_w \qquad [2.37]$$

The minimum values of the hoop stress are obtained at the two opposite points located at $\theta = 0$ and $\theta = \pi$. The diameter joining these two points is parallel to the maximum in situ stress normal to the borehole axis, σ_H. The minimum value of the hoop stress is given by:

$$\sigma_{\theta min} = 3\sigma_h - \sigma_H - P_w \qquad [2.38]$$

The minimum value of the hoop stress given by Eq. [2.38] plays an important role in the analysis of tensile fracture initiation at the borehole wall.

Borehole stability analysis is used to evaluate whether failure will occur at the borehole wall under the given in situ stresses, pore pressure, and borehole pressure. Failure criteria, such as the Mohr–Coulomb criterion, can be invoked to check whether the shear stresses in the borehole wall exceed the shear strength of the rock. If this is the case, breakouts can be induced, schematically shown in Fig. 2.16. The Mohr–Coulomb criterion yields the following value of the borehole pressure at which shear failure commences at the borehole wall, provided the in situ stresses are such that $\sigma_\theta > \sigma_z > \sigma_r$, the far-field horizontal stresses are isotropic, and the pore pressure remains unchanged [1]:

$$P_w = P_p + \frac{2(\sigma_h - P_p) - C_0}{1 + \tan^2\left(\frac{\pi}{4} + \frac{\varphi}{2}\right)} \qquad [2.39]$$

Shear failure occurs when the borehole pressure drops below this value. Other combinations of principal stresses, for example $\sigma_z > \sigma_\theta > \sigma_r$, need to be considered in a full borehole stability analysis in order to obtain the minimum allowed value of the borehole pressure [1]. This value yields the lower

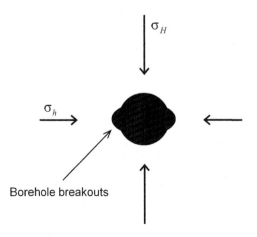

FIGURE 2.16   Borehole breakouts.

bound of the drilling pressure window. Reducing the borehole pressure below this lower bound may result in shear failure, with cavings (pieces of broken rock) falling into the well. Such failure of the hole wall can contribute to solids accumulation at the bottomhole, in the most severe cases resulting in stuck pipe. This failure mode is, however, not directly related to our main subject, the lost circulation.

From the lost-circulation perspective, the *upper* bound of the drilling window is much more important. The upper bound of the drilling window, the *lost-circulation pressure*, is covered in great depth in chapter "Mechanisms and Diagnostics of Lost Circulation." This upper limit can rarely be evaluated from the linear elastic stress analysis alone, and requires that the physics of lost circulation, including the processes of fracture initiation and growth, be taken into account. However, our simple linear elastic analysis performed above allows us a closer look at one particular aspect of lost circulation, namely the *fracture initiation pressure* (FIP). Consider an ideal case of a perfectly intact borehole wall, without any preexisting natural or drilling-induced fractures. The fracture initiation pressure is defined as the value of the borehole pressure at which a fracture is initiated at the intact borehole wall. If such fracture has been initiated, and the fluid pressure is further increased, the fluid will eventually enter the fracture, and the fracture will propagate into the rock in an unstable way. The unstable fracture propagation is initiated when the borehole pressure is equal to the *formation breakdown pressure* (FBP).

As long as the wellbore pressure is below FBP, the fluid is not able to enter the fracture, because the fluid is too thick, and the solid particles in the fluid are too large. In this case, the fracture will stop growing and will stabilize. Thus, *initiation of a fracture is not sufficient to cause lost circulation.*

The fracture initiation pressure can be estimated from the linear elastic analysis presented above. Initiating a fracture in the radial direction requires that the absolute magnitude of the effective hoop stress becomes larger than the tensile strength. Assuming, as a first approximation, that the uniaxial tensile strength of the rock, T_0, can be used as the tensile strength, the criterion for initiation of a radial fracture is given by:

$$\sigma'_\theta \leq -T_0 \qquad [2.40]$$

where the minus in front of T_0 indicates that the effective stress must be tensile and thus negative.

In the case of an impermeable borehole wall, Eqs. [2.38] and [2.40] suggest that a radial fracture is initiated when the borehole pressure exceeds the FIP given by

$$FIP = 3\sigma_h - \sigma_H - P_p + T_0 \qquad [2.41]$$

where P_p is the pore pressure in the near-well area.

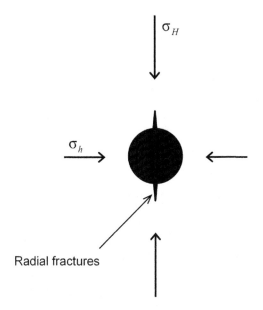

FIGURE 2.17   Radial fractures in the borehole wall.

In the case of a permeable borehole wall, the initiation pressure for a fracture propagating in the radial plane (Fig. 2.17) is given by [1]:

$$\text{FIP} = \frac{2\sigma_h - \alpha \frac{1-2v_d}{1-v_d} P_{p0} + T_0}{2 - \alpha \frac{1-2v_d}{1-v_d}} \qquad [2.42]$$

where P_{p0} is the initial pore pressure before drilling (the far-field pore pressure). The plane of the fracture initiated according to the criteria given by Eqs. [2.41] and [2.42] is perpendicular to the minimum far-field stress, σ_h. Since, in a perfectly circular borehole, the criterion given by Eqs. [2.41] or [2.42] is satisfied simultaneously at two diametrically opposite lines at the borehole wall, the initiated fracture will have two wings, as schematically shown in Fig. 2.17.

Eqs. (2.41) and (2.42) represent the upper and the lower limit of FIP, respectively. Since, in reality, the borehole wall is neither perfectly impermeable nor perfectly pervious to the drilling fluid, the actual value of FIP will be somewhere between these two values.

If the stress acting along the borehole axis is sufficiently small, a fracture can be initiated in the plane perpendicular to the borehole axis, rather than in the radial plane. The criterion for the fracture initiation then is given by

$$\sigma'_z = -T_0 \qquad [2.43]$$

Tensile fracturing governed by Eq. [2.43] may be relevant, for example, for a vertical borehole in a compressional tectonic environment, where σ_v is considerably lower than σ_h and σ_H.

Drilling fluid can have a temperature different from the formation. Similar to the cooling of the reservoir as a whole (Section 2.5), a local cooling of the near-well area by the colder mud induces thermal stresses. In particular, the hoop stress and the axial stress at the borehole wall are changed by [1,13]:

$$\Delta\sigma_\theta^{thermal} = \Delta\sigma_z^{thermal} = \frac{E_d}{1 - \nu_d}\alpha(T_w - T_0) \qquad [2.44]$$

where T_0 is the initial temperature in the near-well area (prior to drilling); T_w is the temperature in the well. Thus, cooling reduces the hoop stress and the axial stress. Therefore, cooling will facilitate initiation of both radial fractures and fractures normal to the borehole axis; ie, the FIP will be lowered.

2.7 MEASUREMENT OF IN SITU STRESSES

Predicting borehole failure and fracturing during drilling requires that rock properties and in situ stresses are known. While a number of techniques are available for measuring rock stresses in civil engineering and mining engineering, this task is considerably more complicated in oil and gas industry because there is not direct access to the rock.

In general, stress measurements should provide information about directions and magnitudes of in situ stresses. In tectonically inactive regions, the vertical direction can be assumed as the direction of σ_1, with reasonable confidence. The magnitude of this stress can then be evaluated by calculating the weight of the overlaying rocks according to Eq. [2.8]. The rock density profile needed in Eq. [2.8] can be obtained, for example, from a density log.

Evaluation of horizontal stresses and their directions requires stress measurements. A number of stress measurement techniques have been developed and used with varying degree of success. Stress measurement methods can be roughly subdivided into core-based methods and in situ methods.

Core-based methods imply that in situ stresses have introduced an "imprint" in the rock that can be assessed by laboratory measurements on the core material and hence be used to deduce the in situ stresses. For example, such an imprint can be the amount and orientation of microcracks created by in situ stresses, the amount of deformation, etc. Examples of core-based methods are the method based on the *anelastic strain recovery* (ASR) and the method based on the *Kaiser effect*. The ASR method makes use of viscoelastic behavior of the rock. When a core is retrieved from the formation, it will tend to assume the shape corresponding to the stress-free state. The deformation (expansion) of the rock has two components: an instantaneous strain recovery caused by Hooke's elasticity and a delayed, gradual strain recovery caused by viscoelastic behavior. The former happens quickly and, in practice, cannot be measured. The latter develops over hours after the core retrieval and can be measured if strain gauges are installed on the core immediately after retrieval. From rock mechanical properties of the rock, and the measured viscoelastic strains, the in situ stress values and directions can be deduced [1].

Another core-based technique makes use of the so-called *Kaiser effect* [14]. The Kaiser effect is observed when a rock is subjected to loading, unloading, and reloading to a higher stress level. If the stress reaches sufficiently high level during the first loading (but still below failure), microcracks are created in the rock. These microcracks give rise to measurable acoustic pulses, the phenomenon known as *acoustic emission*. Acoustic emission can be thought of as a laboratory-scale counterpart of microseismicity observed at a much larger scale in situ. Upon unloading, the rock contains damage—ie, microcracks—created during loading. Thus, information about the first loading is "memorized" in the rock. During a subsequent reloading, all microcracks that could be generated at stresses below the maximum previously applied stress, have already been generated in the first loading. Thus, there will be no new damage, and thus no acoustic emission up until the stress becomes equal to the maximum previously applied stress. When the stress becomes equal to the highest previously applied stress, acoustic emission resumes. The rock has effectively "memorized" the highest previously applied stress and "remembers" it when it is reloaded [15].

If we consider the in situ stress state as the first loading cycle and the laboratory loading as the second, it should be possible to use the Kaiser effect for stress measurements. Practical application of this technique has, however, been limited. The reason is the complicated stress history the rock has experienced in situ. In addition, the stress state in the formation is triaxial with, in general, unknown principal stress directions. The laboratory reloading is uniaxial or triaxial, with the principal stress directions that only accidentally may coincide with the in situ principal stress directions.

The unknown rock history complicates the use of core-based stress measurement methods. Another significant shortcoming of core-based methods is their local nature. The measurement effectively provides information about stresses at one location. Rock mechanical properties are often required to interpret the test results. Since rocks are heterogeneous materials, in terms of mechanical properties, crack distribution, and in situ stress distribution, the results obtained on a rock sample might not be representative for a larger rock mass.

Shortcomings of core-based methods outlined above have led to the development of in situ stress measurements methods. One such method used in oil and gas industry is the extended leakoff test (XLOT). The test is performed in a well, after the casing has been set and cemented. A few meters of the formation is then exposed by drilling below the casing shoe. The formation is then loaded by increasing the pressure in the well. The injection rate is kept constant, and the borehole pressure is recorded (Fig. 2.18). Up to a certain pressure, the *leakoff pressure* (LOP), the curves "pressure versus time" and "pressure versus injected volume" are linear. The LOP is usually interpreted as

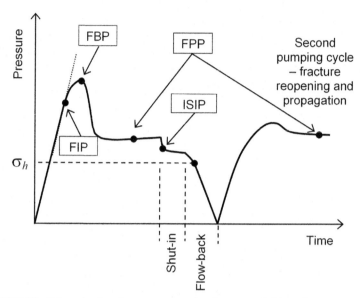

FIGURE 2.18 Schematic plot of pressure versus time in an extended leakoff test (XLOT). *FIP*, fracture initiation pressure; *FBP*, formation breakdown pressure; *FPP*, fracture propagation pressure; *ISIP*, instantaneous shut-in pressure.

the fracture initiation pressure discussed earlier in this chapter, or as the pressure at which the borehole fluid starts to leak through the borehole wall into the exposed formation [2].

After the fracture initiation, the pressure continues to rise until at the FBP, the fracture becomes unstable. The fracture breaks through, and the pressure drops to its value required for fracture propagation (the *fracture propagation pressure*). The test is then continued further, long enough for the fracture front to leave the near-well area of stress concentration—ie, until the fracture length in the radial direction becomes on the order of 5–10 borehole diameters. Then the next stage of the XLOT is performed, namely the shut-in phase. The injection is stopped. This terminates the fracture growth and the fluid flow into the fracture. As a result, the pressure drops to the *instantaneous shut-in pressure* (ISIP). This drop occurs because there is no longer need to create new fracture surface by breaking the rock at the fracture tip, and also because there is no longer frictional pressure drop inside the fracture since the flow has stopped. In earlier interpretation techniques, the ISIP was often used as an estimate of the minimum in situ stress. This interpretation is, however, hardly justified since the fracture is still opened at ISIP. Thus, ISIP is likely to be somewhat higher than σ_h.

After the pressure has reached ISIP, it continues to decrease. The rate of this decrease depends on the leakoff rate through the fracture wall into the formation. If the leakoff occurs sufficiently rapidly or if enough time is allowed for the shut-in phase, the pressure will eventually level off at the *fracture closure pressure* (FCP). The FCP is sometimes interpreted as an estimate of the minimum in situ stress or its upper bound [1].

A more accurate estimate of the minimum in situ stress can be obtained by performing a flowback stage of the test. The flowback stage is performed after the shut-in phase. During the flowback, the fluid is allowed to drain from the fracture back into the well. Monitoring the pressure evolution during the flowback phase reveals that a change of slope in the pressure versus time curve or the pressure versus volume curve occurs. This happens when the fracture has closed, and the fracture faces have come into contact. The stiffness of the system when the faces are in mechanical contact is higher than when the fracture was opened, which gives rise to the change of slope. The pressure at which this change of slope takes place provides probably the most accurate estimate of the minimum in situ stress available from XLOT [16].

Interpretation of XLOT is complicated by many uncertainties present in real downhole conditions. Preexisting and drilling-induced fractures, heterogeneities in the near-well area, plasticity of the rock, and compressibility of the fluid introduce uncertainties into the interpretation [2].

XLOT may provide an estimate of the minimum in situ stress, but gives no clue about the magnitude of other principal stresses. In particular, estimating the intermediate stress is a great challenge. In theory, it should be possible to evaluate the intermediate stress from the fracture initiation pressure, using Eqs. [2.41] or [2.42], if the rock properties, the minimum in situ stress, and the pore pressure are known. In practice, however, the uncertainty involved in identifying the FIP from the pressure curve, and unknown tensile strength of the rock make such evaluation difficult.

Apart from the magnitudes of in situ stresses, their directions are of significance for rock mechanics analyses. In particular, the vertical direction does not necessarily need to be one of the principal stress directions—eg, in the vicinity of faults. The direction of the minimum in situ stress can be estimated if the orientation of the fracture created in an XLOT can be measured. The minimum in situ stress will be normal to the fracture in this case, if the fracture was driven out of the near-well stress concentration zone. Another technique applied in practice to estimate the horizontal stress directions (in the case of a vertical well) makes use of borehole breakouts. As Fig. 2.16 suggests, breakouts develop in the direction of the intermediate stress. Using a caliper log, the location of breakouts on the borehole circumference can be recorded and used to estimate the in situ stress directions. However, uncertainty persists as to whether the observed borehole enlargements are indeed caused by shear failure, or rather by washout and erosion due to the action of the drillstring.

The absence of reliable data about rock properties and in situ stresses at depth are amongst the greatest challenges in predicting lost circulation and designing the treatments. Operators are often reluctant to perform the XLOTs, since fractures created in such tests may weaken the borehole and may thereby increase the risk of losses during subsequent drilling. Quite often, the *formation integrity test* (FIT) or the *leakoff test* (LOT) is performed instead. In FIT, the formation is exposed to a predetermined pressure that is below the LOP. FIT enables the operator to verify that the formation can withstand a given pressure level, but provides no clue as to the in situ stress magnitudes. In an LOT, the bottomhole pressure is increased above the LOP, but the test is stopped soon thereafter, before reaching the FBP.

Typically, LOT, XLOT, and FIT are performed at the casing shoe. The tests thus provide results representative of that specific location only. In complex formations with heterogeneous stress distributions, these results might not be representative for the interval to be drilled. In particular, the fracture gradient in some parts of the next interval may be lower than measured at the casing shoe. "FIT while drilling" can be performed at different locations of an interval while drilling ahead. The method requires, however, that a particular type of managed pressure drilling, namely *constant bottomhole pressure* (CBP) drilling, is used [2]. In FIT while drilling, surface backpressure usually employed in CBP is used to perform the test. As in the conventional FIT, FIT while drilling cannot be used to obtain the magnitude of the minimum in situ stress, but can confirm that a given bottomhole pressure can be applied without inducing losses.

2.8 SUMMARY

Stresses in the reservoir and the near-well area control borehole stability and fracturing during drilling. Depletion usually reduces the total stresses within the reservoir, making it easier to create a fracture if the wellbore pressure becomes too high. In the qualitative analyses performed in this chapter, rock mechanical properties of the reservoir were assumed to be the same as in the surrounding rocks. Only oversimplified shapes of the reservoir were considered—ie, infinite in horizontal directions or ellipsoidal. In reality, rock mechanical properties of the reservoir are different from the overburden, and the overburden itself is not homogeneous. In addition, reservoirs have complex three-dimensional shapes. Moreover, both reservoir rocks and the overburden contain fractures and faults that can affect borehole stability and lost circulation during drilling. In the next chapter, we will have a closer look at natural fractures, a very common source of lost circulation problems.

REFERENCES

[1] Fjær E, Holt RM, Horsrud P, Raaen AM, Risnes R. Petroleum related rock mechanics. 2nd ed. Amsterdam: Elsevier; 2008.

[2] Rehm B, Schubert J, Haghshenas A, Paknejad AS, Hughes J, editors. Managed pressure drilling. Houston, Texas: Gulf Publishing Company; 2008.

[3] Stacey TR, Xianbin Y, Armstrong R, Keyter GJ. New slope stability considerations for deep open pit mines. Jl S Afr Inst Min Metall 2003;(July/August):373—89.

[4]   Willson SM, Edwards S, Heppard PD, Li X, Coltrin G, Chester DK, et al. Wellbore stability challenges in the deep water, Gulf of Mexico: case history examples from the Pompano Field. SPE paper 84266 presented at the SPE Annual Technical Conference and Exhibition held in Denver, Colorado, USA, October 5—8, 2003.

[5]   Fredrich JT, Coblentz D, Fossum AF, Thorne BJ. Stress perturbations adjacent to salt bodies in the deepwater Gulf of Mexico. SPE paper 84554 presented at the SPE Annual Technical Conference and Exhibition held in Denver, Colorado, USA, October 5—8, 2003.

[6]   Fredrich JT, Engler BP, Smith JA, Onyia EC, Tolman DN. Predrill estimation of subsalt fracture gradient: analysis of the Spa prospect to validate nonlinear finite element stress analyses. SPE/IADC paper 105763 presented at the 2007 SPE/IADC Drilling Conference held in Amsterdam, The Netherlands, February 20—22, 2007.

[7]   Zoback MD, Zinke JC. Production-induced normal faulting in the Valhall and Ekofisk oil fields. Pure Appl Geophys 2002;159:403—20.

[8]   Segall P, Fitzgerald SD. A note on induced stress changes in hydrocarbon and geothermal reservoirs. Tectonophysics 1998;289:117—28.

[9]   Orlic B, Heege J, Wassing B. Assessing the integrity of fault- and top seals at CO_2 storage sites. Energy Procedia 2011;4:4798—805.

[10]   Nelson EJ, Hillis RR, Meyer JJ, Mildren SD, Van Nispen D, Briner A. The reservoir stress path and its implications for water-flooding, Champion Southeast field, Brunei. ARMA/USRMS paper 05-775. In: Alaska Rocks 2005, the 40th US Symposium on Rock Mechanics (USRMS): rock mechanics for energy, mineral and infrastructure development in the northern regions, held in Anchorage, Alaska, June 25—29, 2005.

[11]   Santarelli FJ, Tronvoll JT, Svennekjaer M, Skeie H, Henriksen R, Bratli RK. Reservoir stress path: the depletion and the rebound. SPE/ISRM paper 47350 presented at the SPE/ISRM Eurock'98 held in Trondheim, Norway, July 8—10, 1998.

[12]   Vidal-Gilbert S, Tenthorey E, Dewhurst D, Ennis-King J, Van Ruth P, Hillis R. Geo-mechanical analysis of the Naylor Field, Otway Basin, Australia: implications for CO_2 injection and storage. Int J Greenhouse Gas Control 2010;4(5):827—39.

[13]   Morita N, Black AD, Fuh G-F. Theory of lost circulation pressure. In: SPE paper 20409 presented at the 65th Annual Technical Conference and Exhibition of the Society of Petroleum Engineers held in New Orleans, LA, September 23—26, 1990.

[14]   Lavrov A. The Kaiser effect in rocks: principles and stress estimation techniques. Int J Rock Mech Min Sci. 2003;40(2):151—71.

[15]   Pestman BJ, Van Munster JG. An acoustic emission study of damage development and stress-memory effects in sandstone. Int J Rock Mech Min Sci & Geomech Abstr 1996;33(6):585—93.

[16]   Raaen AM, Horsrud P, Kjørholt H, Økland D. Improved routine estimation of the minimum horizontal stress component from extended leak-off tests. Int J Rock Mech Mining Sci 2006;43(1):37—48.

Chapter 3

Natural Fractures in Rocks

Fractures create escape paths for drilling fluid and thereby constitute an important mechanism of lost circulation. In this chapter, basic information about geological and rock mechanical properties of natural fractures is provided (Sections 3.1 and 3.2). Fluid flow in fractures is discussed in Section 3.3, in particular flow of non-Newtonian yield-stress fluids such as drilling fluids. The effect of fracture roughness on fluid flow and particle transport is outlined. The concepts of hydraulic and mechanical aperture are introduced, which are crucial for understanding lost circulation in naturally fractured rocks. The effect of normal and shear stresses on fracture permeability is demonstrated. Implications of particle size distribution on particle transport in real-life rough-walled fractures are discussed (Section 3.4).

3.1 GEOLOGICAL ASPECTS OF NATURAL FRACTURES

A fracture is a discontinuity in a material. Most rocks contain fractures of various sizes, from microcracks at grain level to fractures extending for hundreds of meters in the reservoir. In some reservoir rocks, fractures provide important pathways for reservoir fluids. Production from naturally fractured reservoirs with low matrix permeability, such as chalk or gas-bearing shale, is often only possible because a connected fracture system exists that can deliver the hydrocarbons to the producing well.

Fractures can be classified by their origin as tensile fractures or shear fractures. *Tensile fractures*, often called *joints* in geological literature, are created by tensile failure, also known as mode I fracturing in fracture mechanics (Fig. 3.1A). The relative displacement of the fracture faces is normal to the fracture plane in this mode.

Shear fractures are created by relative displacement of the fracture faces along the fracture plane. This can happen either by shear in the direction normal to the fracture front (mode II, Fig. 3.1B) or in the direction parallel to the fracture front (mode III, Fig. 3.1C).

Mixed-mode fracturing is quite common, too. An example is a combination of mode II and mode III in a shear event. Due to the complex geometry of fractures in rocks, different modes may prevail in different parts of the same fracture.

Tensile (mode I) fractures can undergo shear in their subsequent geological history. Thus, even though fracture surfaces of a tensile fracture might match

Lost Circulation. http://dx.doi.org/10.1016/B978-0-12-803916-8.00003-0

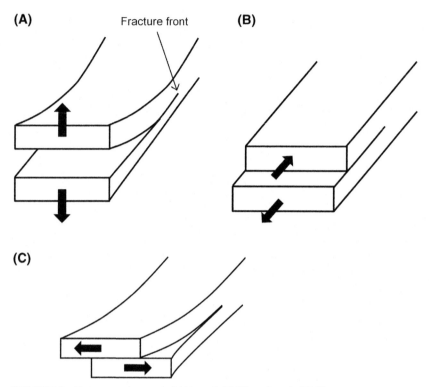

(A) Fracture front **(B)**

(C)

FIGURE 3.1 Fracture modes: mode I (A), mode II (B), and mode III (C).

each other at the time of creation, they may subsequently be displaced and thus not match in present time.

Fracture surfaces are rarely perfectly smooth. Tensile fractures are known to have roughness that reflects the history of fracture propagation. For instance, in sedimentary rocks, roughness may take the form of so-called *hackle plumes*. Grooves and ridges making up a hackle plume diverge in the direction of the fracture growth. In addition, *rib marks* may be created roughly perpendicular to the hackle plume pattern. The ribs indicate the positions of the fracture front at temporary fracture arrests (Fig. 3.2) [1]. Surface roughness of fracture faces can be measured by means of mechanical or laser profilometry [2–4].

Opening of an individual fracture can be quantified by its *mechanical aperture*, w, which is defined as the average distance between the fracture surfaces [3]:

$$w = \frac{1}{N} \sum_{i=1}^{N} w_i \qquad [3.1]$$

where N is the number of locations where the aperture is measured; w_i is the aperture at the ith location. The variable w_i can be called the *local aperture*,

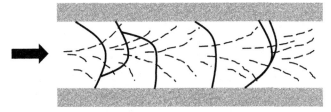

FIGURE 3.2 Surface markings indicating the propagation history of a tensile fracture confined between two layers. The overall propagation direction is indicated by *arrow*. *Solid lines* represent ribs (fracture front positions at temporary fracture arrests). *Dashed lines* show a hackle plume pattern pointing in the direction of the fracture growth. *Based on Twiss RJ, Moores EM. Structural geology. 2nd ed. W.H. Freeman; 2007. 500 p.*

as it quantifies the fracture opening at a particular location. The local nature of w_i distinguishes it from the mechanical aperture, w, and from the hydraulic aperture, w_h, to be introduced later in this chapter. Both mechanical and hydraulic apertures are global average parameters and characterize the fracture as a whole.

Fracture roughness can be quantified by means of the standard deviation of the aperture given by

$$\delta = \sqrt{\frac{1}{N-1} \sum_{i=1}^{N} (w_i - w)^2} \qquad [3.2]$$

Roughness of fracture faces and the aperture distribution depend on the history of the fracture deformation, deposition of minerals transported by fluids flowing inside the fracture, temperature changes, etc. In particular, the deposition of minerals such as calcite or quartz inside a fracture creates new asperities and reduces the permeability of the fracture. In addition to mineral precipitation, gouge (crushed rock) can be deposited in the fracture when asperities are crushed during shear.

Since natural fractures are typically created by large-scale processes caused by, for example, tectonic forces, single fractures are rare. Typically, fractures in sedimentary rocks are arranged in fracture sets (Fig. 3.3). A *fracture set* is a collection of fractures having approximately the same orientation. Spacing between the fractures in a set depends on the geological history and the rock properties. Fractures normal to bedding in sedimentary rocks usually have spacing that increases with the layer thickness to which the fractures are confined.

More than one fracture set are often present in sedimentary rocks. A collection of all fracture sets is known as the *fracture system*. An example of a fracture system consisting of two fracture sets is shown in Fig. 3.3. A fracture system can be unconnected or connected. In the latter case, hydraulic communication can be established between different parts of the fracture

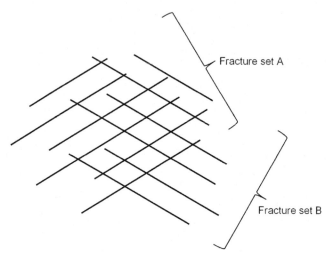

FIGURE 3.3 Fracture system composed of two fracture sets. The fractures are represented by their traces in the plane of the figure.

system, which in this case will represent a *fracture network*. Connectivity of the fracture network is its essential property. In the lost circulation context, it affects how much drilling fluid can be lost if such a network is met during overbalanced drilling. In naturally fractured reservoirs, the availability of a connected fracture system is essential for production, but is detrimental for drilling. An example of a mineralized fracture network in an outcrop rock is shown in Fig. 3.4.

The geometry of rock fractures is determined by the stress history and heterogeneity of the rock. An example of a Z-shaped fracture is shown in Fig. 3.5. Such kinks in fracture trajectories are common and have been observed also in man-induced fractures—eg, in those created during mine-back hydraulic fracturing experiments [5]. In hydraulic fracturing, such Z-shaped fractures are sometimes called *off-balance fractures*. Arguably, they may induce screenouts during proppant injection, since the aperture of the fracture in the inclined part is reduced, thus obstructing particle transport through such fracture [6].

Fracture growth may result in some fractures meeting others. What happens to a fracture when its tip approaches another fracture depends on many factors, such as fracture orientations with respect to in situ stresses and to each other, rock heterogeneity, etc. Some possible scenarios are shown in Fig. 3.6. In Fig. 3.6A, the propagating fracture crosses the preexisting fracture as if there were no obstacle. In Fig. 3.6B, the propagating fracture is offset by the preexisting fracture, which results in a Z-shaped geometry. In Fig. 3.6C, the propagating fracture is arrested. An example of a fracture that might have been created in a way similar to that of Fig. 3.6C is shown in Fig. 3.7. The fracture

FIGURE 3.4 Mineralized fracture network in an outcrop rock in West Greenland. *Photograph by the author. The original is available at: https://www.flickr.com/photos/lavroff/14633235130/.*

indicated by the white arrow was apparently created *after* the other fracture in the figure had already developed. (Otherwise it would be quite unlikely for the latter to pass exactly through the tip of the former.)

Natural fractures can be created in a variety of geological processes. In particular, both tensile and shear fractures are common around faults, increasing the risk of mud losses while drilling in such zones. Tensile fractures around faults are known as *pinnate fractures* (Fig. 3.8) [1].

Fractures of both tensile and shear type can be associated with folding in sedimentary rocks. Some examples are shown in Fig. 3.9: fractures on the convex side of the fold are tensile, while those on the concave side are induced by shear failure.

Fractures may develop during the entire geological history of the rock. In sedimentary rocks, fractures can develop even before the rock becomes

FIGURE 3.5 Z-shaped fracture in an outcrop rock in West Greenland. *Photograph by the author.*

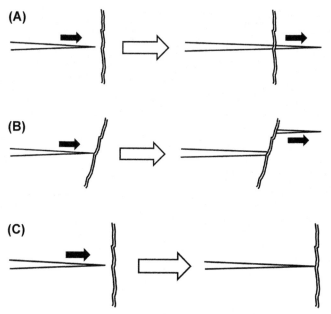

FIGURE 3.6 Fracture propagation in a rock with preexisting fractures. *Full arrows* indicate the direction of fracture propagation. *Hollow arrows* indicate the sequence of events. The three scenarios shown are:

(A) propagating fracture resumes propagation on the other side of the preexisting fracture;
(B) propagating fracture becomes offset by the preexisting fracture;
(C) propagating fracture becomes arrested.

FIGURE 3.7 A T-junction comprising two fractures in an outcrop rock in West Greenland. *Arrow* indicates plausible propagation direction of the fracture until it meets a preexisting fracture and becomes arrested. *Photograph by the author.*

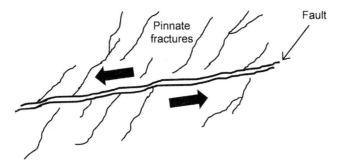

FIGURE 3.8   Pinnate fractures near a fault. The shear direction is indicated by *arrows. Based on Twiss RJ, Moores EM. Structural geology. 2nd ed. W.H. Freeman; 2007. 500 p.*

consolidated. The orientation and other properties of fractures are determined by the in situ stresses and geological conditions that prevailed at the time. Thus, the orientation of fractures currently found in rocks is not necessarily consistent with the *contemporary* in situ stresses.

The shape of individual natural fractures may be highly complex. In an igneous rock—eg, granite—isometric, roughly circular, or elliptical fractures can be found. Sedimentary rocks often have fractures oriented normal to bedding and confined between the adjacent layers of another rock.

Characterization of a fracture system requires that the following data be collected for each of the fracture sets:

- orientation;
- spacing between fractures in the set;
- aperture and roughness of the fractures;
- length of the fractures.

In addition, connectivity between fractures in the fracture system is important.

While the above-listed data can be obtained in an outcrop rock, the task of fracture characterization becomes more difficult at depth. Microresistivity imaging tools reveal locations and spacing of fractures [7]. Combined use of

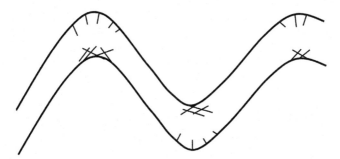

FIGURE 3.9   Fractures associated with folding.

acoustic and microresistivity imaging tools enables discrimination between open and closed fractures [8]. However, image log devices usually do not have sufficient resolution to measure the fracture aperture and roughness of the fracture faces. In addition, a borehole only provides a local view of the fracture system. This last problem persists also in the case of an outcrop rock: only traces of fractures on the rock surface are observable. This gives little clue about three-dimensional fracture extent and connectivity inside the rock mass.

The difficulty of fracture characterization is one of the reasons why it is so difficult to predict the occurrence and potential severity of mud losses in naturally fractured formations.

3.2 ROCK MECHANICAL PROPERTIES OF NATURAL FRACTURES

Lost circulation is a complex, coupled hydromechanical process. Therefore, understanding fracture behavior under normal and shear stresses is important for understanding the mechanisms of lost circulation.

Consider normal deformation (opening/closing) of a planar fracture under normal stress. Fracture opening/closing is controlled by the difference between the normal stress acting in the rock in the direction normal to the fracture plane and the fluid pressure inside the fracture, which we for now assume to be constant. This setup is shown in Fig. 3.10. In the left panel, the total normal stress is greater than the fluid pressure inside the fracture, and the fracture is mechanically closed, ie, there is contact between the fracture faces. Force is transmitted through the contact spots. If the fluid pressure inside the fracture is increased while σ_n is kept constant, the contact forces decrease, and the contacts gradually open up. This process will continue until, eventually, the fracture opens completely, and there are no longer any contact spots. This happens when the *net pressure* inside the fracture, defined as $P^{(net)} = P_p - \sigma_n$, becomes positive (right panel in Fig. 3.10). The net pressure has the same magnitude as and a different sign than the effective normal stress if the Biot effective stress coefficient, α, is equal to 1, as is commonly assumed in failure criteria.

Suppose now that instead of increasing the fluid pressure we would gradually decrease it, starting again from the state shown in the left panel of Fig. 3.10 and keeping σ_n constant. In this case, the fracture would close further; ie, its aperture defined by Eq. [3.1] would decrease. Because of the roughness of the fracture faces, more and more asperities will come into contact, and the area of existing and new contacts will gradually increase. This will increase the fracture's normal stiffness: it will take more and more pore pressure reduction to decrease the aperture by the same amount as the fracture keeps closing. If the applied normal stress were sufficiently high and the rock were linear elastic, then at some point the fracture would close completely; ie,

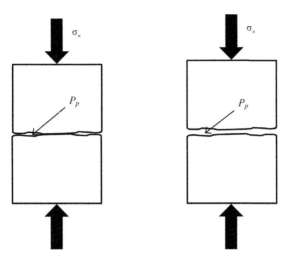

FIGURE 3.10 Fracture under normal stress. P_p is the fluid pressure inside the fracture, and σ_n is the total normal stress acting in the rock in the direction normal to fracture. Left panel: $P_p < \sigma_n$. Right panel: $P_p > \sigma_n$.

there would be no open spots remaining in the fracture. However, in reality this might require the normal stress to be so high that the bulk rock will break before the state of a completely closed fracture is reached. Whether such a completely closed state can be achieved in practice depends on:

- the strength of the rock;
- the stiffness of the rock;
- the height of asperities, characterized, for example, by the standard deviation of the aperture, δ.

The process of normal deformation described above results in the stress–displacement curve shown in Fig. 3.11. At zero effective normal stress, the fracture faces touch each other, but the contact spots do not bear any load. As the stress is increased, the fracture closes, and the normal stiffness (the slope of the curve in Fig. 3.11) increases. The curve has a vertical asymptote because it is only possible to reduce the aperture until the fracture closes completely. The schematic plot in Fig. 3.11 is based on laboratory experiments and numerical modeling [9,10]. The shape of the curve is solely due to the nonmatching rough fracture faces. If the fracture faces were perfectly matching with no gouge in between and no other obstacles in the fracture, the deformation curve would look like the one shown in Fig. 3.12.

Reduction of the mechanical aperture with increasing normal stress (or decreasing fluid pressure in the fracture) brings about a reduction in the fracture permeability. The effect of normal loading on the fracture permeability is discussed in Section 3.3.

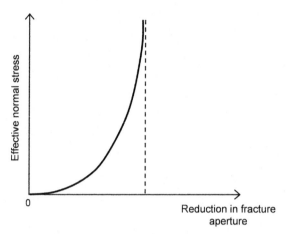

FIGURE 3.11 Normal deformation of a rough-walled fracture. At zero effective stress, fracture faces touch each other but bear no load.

Consider now a fracture subject to shear. A schematic view of an experimental setup for testing shear deformation of a fracture is shown in Fig. 3.13. As the shear displacement is increased, the shear stress increases up until it reaches a peak (Fig. 3.14). The peak value depends on the rock properties and the fracture roughness, as well as on the applied effective normal stress, $(\sigma_n - P_p)$, where P_p is the fluid pressure inside the fracture. The shear displacement at which the shear stress peaks is typically on the order of a millimeter or a fraction of a millimeter in laboratory tests [9,11]. After the peak, the shear stress drops to its residual value, and the sliding continues.

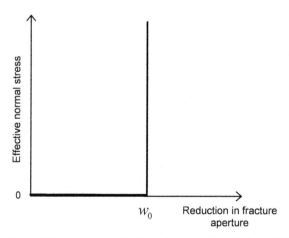

FIGURE 3.12 Normal deformation of an imaginary fracture with perfectly matching or perfectly smooth faces. At zero effective stress, fracture faces do not touch each other and are separated by a distance w_0.

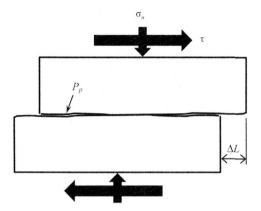

FIGURE 3.13   Fracture under shear stress, τ. P_p is the fluid pressure inside the fracture, and σ_n is the total normal stress acting in the rock in the direction normal to fracture ($P_p < \sigma_n$).

The applied effective normal stress affects the slope of the initial prepeak part of the shear stress versus shear displacement curve in Fig. 3.14. The slope increases with normal stress, the fracture becoming stiffer. Moreover, the normal stress has a profound effect on both the peak shear stress and the residual shear stress: both increase with the effective normal stress, ($\sigma_n - P_p$).

During shear deformation, old contact spots disappear, and new contact spots are created. At low normal stresses, asperities climb over each other, and the fracture aperture, w, increases as is schematically shown in Fig. 3.15. This effect is known as *dilatancy* and can be quantified by the *dilatancy angle*, ψ.

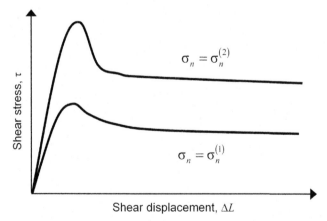

FIGURE 3.14   Fracture deformation under increasing shear stress and constant normal stress σ_n. The *upper curve* corresponds to a higher normal stress, $\sigma_n^{(2)} > \sigma_n^{(1)}$. The fluid pressure inside the fracture is the same for both curves. *Based on Esaki T, Du S, Mitani Y, Ikusada K, Jing L. Development of a shear-flow test apparatus and determination of coupled properties for a single rock joint. Int J Rock Mech Min Sci 1999;36(5):641–50.*

FIGURE 3.15   Fracture deformation under shear stress, τ, at low normal stress, σ_n. Left panel: initial configuration. Right panel: after slip commences, asperities slide over each other, and the fracture dilates (aperture increases).

The increase in the mechanical aperture, Δw, caused by the shear displacement ΔL can be approximated, at the initial stages of sliding, by [12]:

$$\Delta w = \Delta L \tan \psi \qquad [3.3]$$

At low normal stresses, the dilatancy angle is positive; ie, the fracture opens by shear displacement. As the normal stress increases, the dilatancy angle decreases [12]. Eventually, at a sufficiently high normal stress, the forces exerted by asperities on one another will be sufficient to break the asperities (Fig. 3.16). As a result, fracture dilation switches to fracture contraction at high normal stresses, the dilatancy angle eventually becoming negative [9]. The destruction of asperities may further contribute to the reduction of the fracture permeability by accumulation of *gouge* (crushed material) inside the fracture. The effect of the normal stress on dilatancy under shear displacement is illustrated in Fig. 3.17.

Several empirical models have been proposed in rock mechanics to describe the gradual decrease of the dilatancy angle with the effective normal stress, for instance [13]:

$$\psi = \frac{1}{2} \mathrm{JRC} \log \left(\frac{\mathrm{JCS}}{\sigma_n'} \right) \qquad [3.4]$$

where JRC is the joint roughness coefficient, and JCS is the joint wall compression stiffness. According to Eq. [3.4], the dilatancy angle depends on the fracture roughness (JRC), the rock properties (JCS), and the applied effective normal stress. Eq. [3.4] accounts for the transition from dilation to contraction as the effective normal stress, σ_n', increases.

It should now be obvious that fracture behavior under varying normal stresses, shear stresses, and fluid pressure is quite complex and depends on the fracture properties (roughness), the mechanical properties of the rock (elastic moduli, strength), and the stress level (depth). This is why it is

FIGURE 3.16   Fracture deformation under shear stress, τ, at high normal stress, σ_n. Left panel: initial configuration. Right panel: after slip commences, asperities destroy each other, and the fracture closes (aperture decreases).

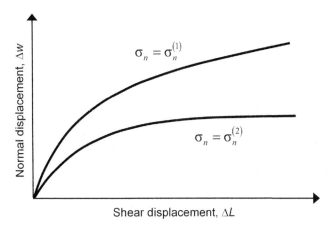

FIGURE 3.17 Normal displacement caused by shear displacement at constant normal stress σ_n. The *lower curve* corresponds to higher normal stress, $\sigma_n^{(2)} > \sigma_n^{(1)}$. The fluid pressure inside the fracture is the same for both curves. *Based on Esaki T, Du S, Mitani Y, Ikusada K, Jing L. Development of a shear-flow test apparatus and determination of coupled properties for a single rock joint. Int J Rock Mech Min Sci 1999;36(5):641–50.*

inherently difficult to predict fracture behavior in situ without an extensive characterization of fractures, which, as discussed in Section 3.1, is rarely achievable. Uncertainties and the lack of data about fracture properties complicate the choice, design, and implementation of lost circulation treatments in fractured rocks.

3.3 FLUID FLOW IN FRACTURES

We begin our discussion of fluid flow in fractures by considering an idealized model of a conduit having parallel smooth walls and a constant aperture, w. The conduit has a rectangular shape in the flow plane (Fig. 3.18).

Let a pressure difference be applied between the left-hand and the right-hand sides of the conduit. In the case of a laminar (Poiseuille) flow of a Newtonian fluid, the flow rate per unit length in the direction normal to flow is given by:

$$q = -\frac{w^3}{12\mu}\frac{P_{p2} - P_{p1}}{L_x} \qquad [3.5]$$

where P_{p1} and P_{p2} are the fluid pressures applied at the two opposite sides of the fracture (Fig. 3.18); L_x is the distance between those sides—ie, the length of the conduit in the direction of flow; μ is the dynamic viscosity of the fluid. Eq. [3.5] represents Darcy's law for the conduit, with the permeability thus given by $w^2/12$.

Real fractures have rough faces, and their aperture is defined by its mean value (the mechanical aperture) and its standard deviation (Eqs. [3.1] and [3.2], respectively). As a result, while the flow in Fig. 3.18 is unidirectional, the flow in a real fracture would be tortuous: to cross from one side to the other, the fluid would have to go around asperities protruding from the fracture faces. At some spots, the fracture might be closed completely because asperities on the opposite faces touch each other or because of gouge deposited in the fracture. As a result, streamlines and pathlines are not straight in real fractures, and are longer than in a smooth-walled conduit. Effectively, this means that the same pressure gradient produces lower flow rate in a rough-walled fracture than in a smooth-walled conduit.

Hydraulic aperture of a rough-walled fracture is defined as the aperture of a smooth-walled conduit that would produce the same flow rate under a given applied pressure gradient as the rough-walled fracture. Hydraulic aperture of a fracture, w_h, is lower than the mechanical aperture, w, sometimes by a factor of 5−10. Permeability of a rough-walled fracture is given by

$$k = \frac{w_h^2}{12} \qquad [3.6]$$

For a conduit with smooth walls, $w_h = w$.

Hydraulic aperture is the aperture that the fluid can "see" as it flows in the fracture. Hydraulic aperture decreases as the standard deviation of the aperture distribution, δ, increases. A semianalytical approximation for w_h as a function of w and δ was proposed by Zimmerman et al. [14]:

$$w_h = w\left\{ \frac{\left[1 + \left(3\delta^2/w^2\right)\right]\left[1 - \left(2\delta^2/w^2\right)\right]^{5/2}}{1 + \left(\delta^2/w^2\right)} \right\}^{1/6} \qquad [3.7]$$

The approximations of w_h given by Eq. [3.7] is plotted in Fig. 3.19. It is evident from Fig. 3.19 that the hydraulic aperture is close to the mechanical aperture when roughness is small. As roughness increases, the hydraulic aperture decreases rapidly. Note that a drop in hydraulic aperture by a factor of three would mean that the fracture permeability decreases by a factor of 9. This illuminates the effect roughness has on the permeability of natural and induced fractures in rocks.

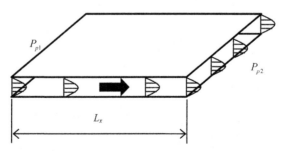

FIGURE 3.18   Unidirectional flow in a conduit with smooth parallel walls. Flow direction is indicated by *arrow*; $P_{p1} > P_{p2}$. Parabolic velocity profiles are characteristic of laminar (Poiseuille) flow of a Newtonian fluid.

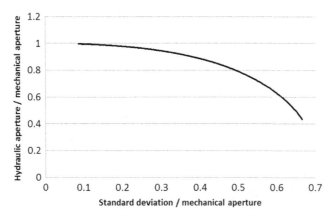

FIGURE 3.19 Normalized hydraulic aperture, w_h/w, as a function of normalized standard deviation of the aperture, δ/w, based on the semianalytical Eq. [3.7].

Fig. 3.19 suggests that when a fracture is subjected to a normal stress, the decrease of its hydraulic aperture will be even greater than the reduction of the mechanical aperture. This is illustrated in Fig. 3.20, based on the experimental findings of Chen et al. [12].

Numerous laboratory experiments have demonstrated the profound effect of the normal stress on the fracture permeability. The effect is nonlinear, and there is hysteresis as shown in Fig. 3.21. The first crucial observation about

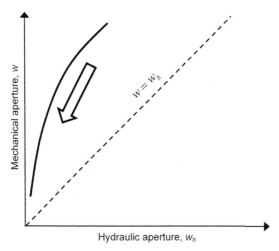

FIGURE 3.20 Mechanical aperture as a function of hydraulic aperture during normal loading of a rough-walled fracture (*solid line*). Both apertures decrease as the normal stress increases. *Dashed line* shows a hypothetical case of hydraulic and mechanical apertures being equal. *Arrow* indicates an increase of the normal stress. *Based on Chen Z, Narayan SP, Yang Z, Rahman SS. An experimental investigation of hydraulic behaviour of fractures and joints in granitic rock. Int J Rock Mech Min Sci 2000;37(7):1061–71.*

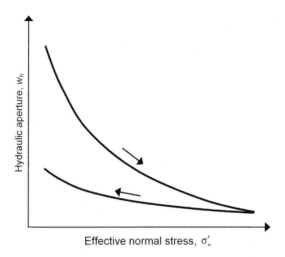

FIGURE 3.21 Hysteresis of hydraulic aperture under normal loading/unloading. *Based on Gutierrez M, Øino LE, Nygård R. Stress-dependent permeability of a de-mineralised fracture in shale. Mar Petrol Geol 2000;17(8):895–907.*

Fig. 3.21 is that the hydraulic aperture, and thus the fracture permeability, does not become zero under normal stress, as long as there remains a connected flow path inside the fracture. It means that even though fracture faces touch each other, the flow of a Newtonian fluid is still possible, and it takes a substantial effort to further reduce the fracture permeability at higher stresses (note the changing slope of the curve in Fig. 3.21). This means that even though a fracture at depth might appear "mechanically closed"—ie, its faces touching each other at many locations—the fracture can still be open for flow; ie, it can be "hydraulically opened."

This distinction was recognized, for example, in Ref. [9]. Furthermore, as pointed out in Ref. [9], it is not always possible to reduce the fracture permeability to zero by simply increasing the normal stress. Zero fracture permeability means that there is no connected flow path from one side of the fracture to the other. Even though eliminating all such pathways by increasing the normal stress is in theory possible, the bulk rock might fail before the required stress level is reached.

The behavior of hydraulic aperture under normal stress in shale can be described, for instance, by [9]:

$$w_h = w_{h0} \exp\left(-0.5C\sigma'_n\right) \qquad [3.8]$$

with the fracture permeability thus given by

$$k = k_0 \exp\left(-C\sigma'_n\right) \qquad [3.9]$$

C being a fitting parameter, equal to 0.27 for Kimmeridge shale [9]; w_{h0} and k_0 being, respectively, the hydraulic aperture and the fracture permeability at

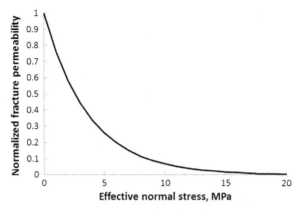

FIGURE 3.22 Hydraulic aperture normalized by its initial value as a function of the applied effective normal stress.

the beginning of loading. The dimension of σ'_n in Eqs. [3.8] and [3.9] is MPa. The normalized fracture permeability, k/k_0, given by Eq. [3.9], is plotted in Fig. 3.22. The exponential dependence of the fracture permeability on the fluid pressure following from Eq. [3.9] is used in reservoir engineering to account for permeability alterations in fractured reservoirs caused by depletion [15,16]. Empirical and semiempirical relationships between the fracture permeability and the normal effective stress were reviewed in Ref. [17].

Hysteresis evident in Fig. 3.21 indicates that asperities undergo irreversible (plastic) deformation during normal loading. Upon unloading, the fracture roughness differs from the one that existed before the loading. The fracture roughness thus bears an imprint of the stress history. This is similar to the so-called stress memory effects known in rocks and man-made materials [18,19].

Laboratory studies of fracture permeability under normal stress are usually performed using setups in which fluid flows between two parallel rough surfaces. Thus, the possible effect of fracture mineralization is neglected. In reality, fractures may be so heavily mineralized that it will change the flow pattern from two-dimensional planar flow into one-dimensional channel flow. For instance, a detailed study of fractures in the Khuff Formation (Saudi Arabia) has shown that 30% of fractures are mineralized, and 46% of fractures have channel-type permeability (Fig. 3.23B) [20]. Channel-type permeability may also be induced by shear displacement of fractures that originally had matching surfaces (Fig. 3.23A) [20]. Permeability of channel-type fractures is less stress-sensitive than that of simple "parallel-plate" fractures commonly used in laboratory experiments. The minerals deposited inside the fracture keep it opened when the effective normal stress increases.

Evolution of the fracture permeability under shear displacement depends on whether the fracture undergoes dilatancy or contraction. Laboratory experiments on shale suggest that, if the normal stress is sufficiently low, the

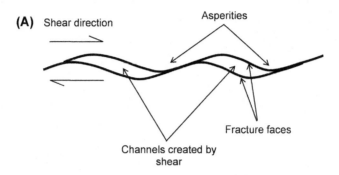

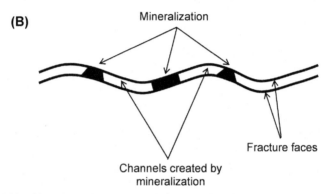

FIGURE 3.23   Channel-type fracture permeability created by shear displacement (A) or mineralization (B). The flow direction is normal to page. *Based on Ameen MS. Fracture and in-situ stress patterns and impact on performance in the Khuff structural prospects, eastern offshore Saudi Arabia. Mar Petrol Geol 2014;50:166−84.*

fracture opens up during the shear, and its permeability increases. If the normal stress is sufficiently high, the fracture aperture decreases under shear displacement, and the fracture permeability decreases. Breakage of asperities and accumulation of gouge inside the fracture additionally contribute to the permeability reduction. Broken pieces of rock would block the pathways for the fluid, effectively reducing the fracture permeability [9].

During shear displacement under low normal stress, the fracture permeability increases and then levels off (Fig. 3.24) [11]. This increase is due to dilatancy and can be by one to two orders of magnitude. The maximum value of permeability achievable during sliding depends on the maximum height of asperities.

During reverse shearing, the fracture permeability declines only slightly, and significant permeability still exists when the fracture faces return to their initial position (Fig. 3.24). The reduction in permeability during reverse shearing is larger under higher applied normal stress, which is attributed to fracture becoming plugged by the gouge produced during shear [11].

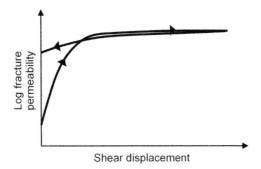

FIGURE 3.24   Log fracture permeability versus shear displacement during forward and reverse shearing under constant applied normal stress. *Based on Esaki T, Du S, Mitani Y, Ikusada K, Jing L. Development of a shear-flow test apparatus and determination of coupled properties for a single rock joint. Int J Rock Mech Min Sci 1999;36(5):641–50.*

The hysteresis (irreversibility) exhibited by the fracture permeability under shear displacement is similar to the irreversibility observed during cyclic normal loading.

Most of our knowledge about fracture permeability under stress is based on experiments performed with hard crystalline rocks that do not exhibit swelling or self-sealing. Studies of stress-dependent permeability of rough-walled fractures in sedimentary rocks, such as shale, are quite rare. One such study was reported in Ref. [9]. Some studies of the fracture permeability under normal stress and shear displacement, performed on Opalinus Clay [21], made use of artificial fractures with idealized smooth surfaces, which unfortunately renders their results largely of academic interest.

Shear displacement can induce anisotropy in the flow patterns inside the fracture. In particular, in a fracture with initially matching faces, the flow after shear displacement becomes more tortuous when the flow is parallel with the shear direction. The flow pattern normal to the shear direction is more regular than the flow pattern parallel to shear. This is due to the channels of increased permeability oriented approximately perpendicular to the shear direction (Fig. 3.23A). The flow normal to shear thus becomes more channelized. This shear-induced preferential channelization can bring about a substantial anisotropy of the fracture permeability. In particular, numerical simulations suggest that the fracture permeability normal to shear can be two to three orders of magnitude larger than the fracture permeability parallel to shear [10]. Anisotropic aperture patterns induced by shear displacement have a bearing on the particle transport in such fractures as will be discussed in Section 3.4.

The majority of experiments on fluid flow in fractures have been performed with Newtonian fluids, typically water. Even though such experiments provide valuable insights into the behavior of fracture permeability as a function of fracture morphology and stress, real drilling fluids are non-Newtonian.

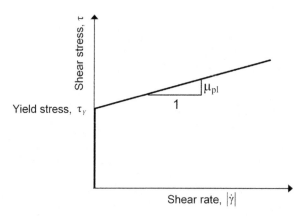

FIGURE 3.25   Shear stress versus shear rate for a Bingham fluid.

Non-Newtonian rheology has a bearing on what happens when a fracture is intersected during drilling. A brief overview of available knowledge on non-Newtonian fluid flow in fractures is therefore provided next.

Typical drilling fluids have yield-stress rheology, which is often approximated in practice in the simplest possible way, namely with the *Bingham model*. The dependency of the shear stress on the shear rate in a simple shear flow is shown in Fig. 3.25 for a Bingham fluid. The shear stress needs to overcome a threshold value τ_Y, called *yield stress*, in order for the fluid to flow. In those domains where the shear stress is below τ_Y, the fluid moves as a solid plug. At shear stresses greater than τ_Y, the Bingham material behaves like a fluid. The slope of the shear stress versus shear rate curve in Fig. 3.25 is called *plastic viscosity*, μ_{pl}.

Let us have a look at how a Bingham fluid flows in a fracture. We start with a laminar flow in a conduit with parallel smooth walls. Due to symmetry, the shear rate must be zero at the center plane of such conduit (the plane equidistant from the walls). Hence, the shear stress in the fluid is zero at the center plane. Therefore, the shear stress is not sufficient to overcome the yield stress near the center plane. As a result, two flow domains coexist in a Bingham fluid flowing between parallel smooth walls: the shear domain near the walls, where the shear stress is sufficiently high to overcome the yield stress, and the unyielded domain of a fluid moving as a solid plug in the middle of the conduit (Fig. 3.26). If the pressure gradient is reduced, the domain occupied by the solid plug expands until, at the pressure gradient equal to $2\tau_Y/w$, it occupies the entire cross section of the conduit, and the flow stops.

Real fractures have rough surfaces. Research on the flow of yield-stress fluids in rough-walled conduits is scarce. Some hints about Bingham fluid flow in fractures can be obtained from analytical and numerical solutions for such fluids in conduits with regular restrictions and expansions. In particular, a unidirectional flow in a conduit with a sinusoidal variation of the aperture

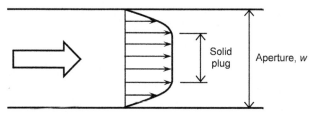

FIGURE 3.26 Schematic view of velocity distribution when a Bingham fluid flows between parallel smooth walls.

results in the solid plug having different velocity in different parts of the conduit: the velocity decreases in expansions and increases in constrictions. Since the plug is a solid material, such a velocity variation implies that the plug must break. It was shown analytically by Frigaard and Ryan that fragmentation of the plug occurs when the variation (standard deviation) of the aperture becomes sufficiently large [22]. This means that continuous solid plugs probably do not exist when the drilling fluid flows in real, rough-walled rock fractures [23].

Another hint about Bingham fluid flow in a rough-walled fracture can be obtained by considering extrusion flows. Fig. 3.27 shows what happens when a yield-stress fluid flows in a conduit with an abrupt expansion [24]. In such flow, the plug must break at the location of the expansion since its velocity is different before and after passing the expansion. In addition, zones of unyielded material develop near the walls right behind the expansion. The shear stress is not sufficient to induce flow in these "dead zones." The size of the "dead zones" increases with the *Bingham number*, Bn, which is defined as

$$\text{Bn} = \frac{\tau_Y D}{\mu_{pl} u} \qquad [3.10]$$

where D is the (average) diameter of the conduit and u is the (average) fluid velocity.

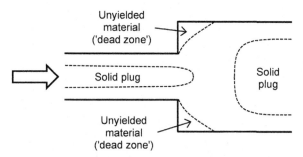

FIGURE 3.27 Flow of a yield-stress fluid in a conduit with an expansion. The *arrow* indicates the flow direction. *Based on Mitsoulis E, Huilgol RR. Entry flows of Bingham plastics in expansions. J Non-Newtonian Fluid Mech 2004;122:45–54.*

In addition to being a function of Bn, the size of the "dead zones" increases with the Reynolds number, Re. At higher Re, the inertial effects carry the fluid farther away from the expansion, and the size of the "dead zones" increases.

Further insight into the flow of a drilling fluid in a rough-walled fracture can be obtained if one considers the apparent viscosity of the fluid. The *apparent viscosity*, μ_{app}, is defined as the ratio of the shear stress to the shear rate. For a Bingham fluid (Fig. 3.25), μ_{app} is given by:

$$\mu_{app} = \mu_{pl} + \frac{\tau_Y}{|\dot{\gamma}|} \qquad [3.11]$$

where $|\dot{\gamma}|$ is the shear rate. As $|\dot{\gamma}| \to 0$, the apparent viscosity becomes infinite, which is signified by the vertical leftmost part of the curve in Fig. 3.25. As the shear rate increases, the apparent viscosity of a Bingham fluid decreases and approaches μ_{pl} as $|\dot{\gamma}| \to \infty$. The fluid thus shows a shear-thinning behavior. When a shear-thinning fluid flows in a rough-walled fracture, the hydraulic aperture that the fluid can "see" is smaller than in the case of a Newtonian fluid flowing in the same fracture [25]. In addition, when a shear-thinning fluid flows around an asperity or a contact spot between the fracture faces, "dead zones" of zero or very low fluid velocity arise as discussed above. This results in channelization of the fluid through the fracture as it finds the path of least hydraulic resistance and streams through it [26]. This behavior will have a bearing on the particle transport, such as transport of lost circulation materials, discussed in the next section.

3.4 PARTICLE TRANSPORT IN FRACTURES

Application of particulates and fibers to prevent or cure lost circulation requires some knowledge of how particles are transported by the drilling fluid in fractured and porous media.

Extensive experimental studies of particle transport in fractured and porous media have been performed in contaminant hydrology. The main focus in those studies is on transport of colloidal particles (tracers). The particle size is typically on the order of $0.1-10$ μm in those experiments. Even though the particle size in typical lost circulation treatments is larger, the results from colloidal transport in fractures provide some interesting insights.

For a fracture of a given aperture and roughness, there exists a particle size, D_*, that is transported most easily. Particles larger than D_* experience settling due to gravity and retention in small-aperture regions of the fracture (bridging). Particles smaller than D_* may penetrate into the porous matrix, may experience diffusion inside the fracture, or may stick to the fracture walls [27].

Particles in horizontal fractures can be transported by rolling on the lower fracture face. Asperities increase the retention of particles. Particles larger than 0.4 of the fracture aperture bridge the fracture, while smaller particles are transported through the fracture [28].

These experimental results, albeit obtained for colloidal particles, demonstrate how important it is to have information about fracture apertures when choosing the particle size that can bridge and seal fractures in lost circulation treatments. The lack of such information may be compensated, at least to some extent, by using a broader *particle size distribution* (PSD).

Transport of colloidal particles in fractures is often characterized by *breakthrough curves*. Consider as an example a laboratory experiment with a rectangular fracture (Fig. 3.28A). Pressure gradient is applied between its sides located at $x = 0$ and $x = L_x$. Tracer is injected at the inlet ($x = 0$), with the constant concentration, C_0. A breakthrough curve illustrates the tracer

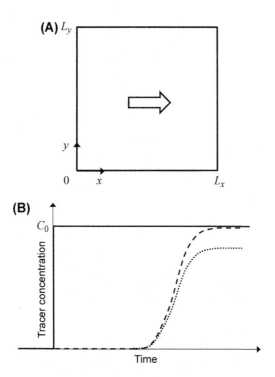

FIGURE 3.28   Flow and transport in a laboratory experiment performed with a rectangular fracture.
(A) In-plane view of the fracture. *Arrow* shows the overall flow direction.
(B) Breakthrough curves:
- *solid line*: injected particle concentration at the inlet ($x = 0$);
- *dashed line*: outlet concentration of particles having size D_*;
- *dotted line*: outlet concentration of particles smaller or larger than D_*.

Based on Koyama T, Li B, Jiang Y, Jing L. Numerical simulations for the effects of normal loading on particle transport in rock fractures during shear. Int J Rock Mech Min Sci 2008; 45(8):1403–19.

concentration at the outlet as a function of time. Two examples are shown in Fig. 3.28B (dashed line and dotted line). First thing to notice about the breakthrough curves in Fig. 3.28B is that the increase in the concentration at the outlet is gradual, contrary to the concentration at the inlet represented by the solid line in Fig. 3.28B. This gradual increase is due to hydrodynamic dispersion cause by the flow tortuosity. If all particles were following the same path, the increase in the concentration at the outlet would be steplike, similar to the concentration at the inlet. Asperities make different particles follow different paths. As a result, particles introduced at different locations along the left-hand side of the fracture arrive at the right-hand side at different times. This brings about the dispersion observed in Fig. 3.28B. Note that this dispersion has nothing to do with molecular dispersion and takes place with particles of any size [29].

The effect of particle size on their transport is evident in Fig. 3.28B. Virtually all particles of size D_* reach the outlet. As a result, the concentration of such particles at the outlet gradually approaches the concentration introduced at the inlet, C_0 (dashed line in Fig. 3.28B). Particles having the diameter above or below D_* are retained in the fracture. As a result, their concentration at the outlet levels off at some value below the concentration introduced at the inlet (dotted line in Fig. 3.28b). This phenomenon was designated as "size segregation" by Cumbie and McKay [27]: the fracture acts as a filter preferentially letting through particles within a certain size range. This has a bearing on how particulate lost circulation materials (LCMs) are placed in fractures. If LCM particles are too big, they will not be able to enter the fracture, and will be deposited at the borehole wall. At that location, the LCM bridge can be easily destroyed by the drillstring action. If particles are too small, they may enter the pore space and thus will not contribute to bridging and sealing the fracture. This is the reason why it is recommended to use a broad PSD in lost circulation treatments: with a broad PSD, it is more likely that there will be some particles able to seal the fracture.

Numerical modeling provides valuable insights into how fracture roughness affects the fluid flow and transport of colloidal particles (tracers). An example is shown in Fig. 3.29. A square fracture has in-plane dimensions of 64 cm × 64 cm. The fracture has rough walls; the aperture distribution is shown in Fig. 3.29A. In the numerical simulation, the fracture is punched by a circular well in the center. A quasi-steady flow from the well into the fracture is established, which produces the fluid velocity distribution shown in Fig. 3.29B. It is evident from Fig. 3.29B that roughness affects the flow pattern significantly: in a fracture with smooth parallel walls, the flow pattern would be perfectly radial. The velocity plotted in Fig. 3.29B is the local fluid velocity averaged over the fracture aperture.

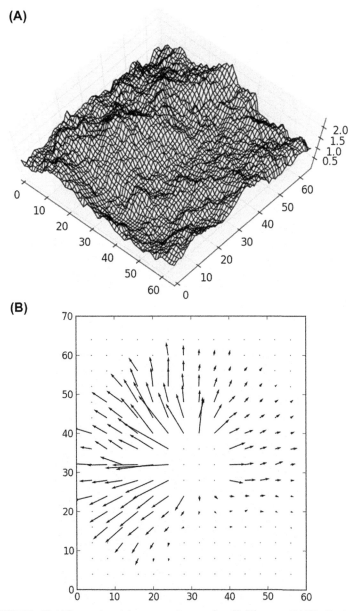

FIGURE 3.29 Fluid flow and particle transport in a rough-walled fracture: (A) Distribution of the fracture aperture; in-plane fracture dimensions: 64 cm × 64 cm; aperture in mm. (B) Fluid velocity (averaged locally along the fracture aperture) when fluid is injected from a circular well in the middle of the fracture. Panels (C), (D), and (E) show subsequent positions of the particle concentration front. Particles are injected from the well into the fracture and are transported by the quasi-steady flow. Noncircular shape of the particle concentration front is due to fracture roughness.

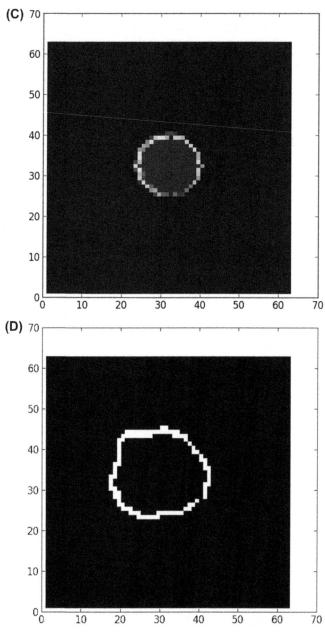

FIGURE 3.29 Cont'd.

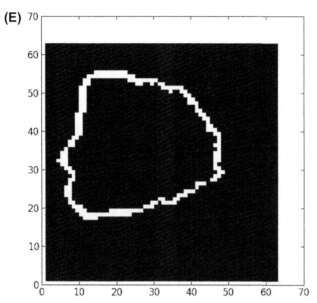

(E)

FIGURE 3.29 Cont'd.

After the flow is established in Fig. 3.29B, particles are introduced at the well, and are transported through the fracture. The advection equation governing the tracer transport by incompressible fluid is given by

$$\frac{\partial C}{\partial t} + u\frac{\partial C}{\partial x} + v\frac{\partial C}{\partial y} = 0 \qquad [3.12]$$

where C is the tracer concentration; u, v are the fluid velocity components in the fracture plane, along the x, y coordinate axes, respectively (see Fig. 3.28A); t is time.

The plots in Fig. 3.29C−E show the particle concentration front at three subsequent times. If the fracture had no roughness, the concentration front would remain circular at all times. Roughness makes the particle front take on a complex, irregular shape governed by the fracture landscape (the asperities).

Normal or shear loading of a fracture affects the fracture aperture distribution and, hence, has a bearing on the particle transport. Normal compression brings about more tortuous transport paths because particles have to travel around larger (and more abundant) contact spots as the stress increases [10]. Thus, particles travel slower, and more particles are retained by asperities and get stuck in the fracture. This improves the bridging and sealing effect and affects the breakthrough curves as shown in Fig. 3.30.

Shear displacement facilitates the flow and thereby the particle transport provided that the fracture experiences dilatancy during shear, ie, it opens up.

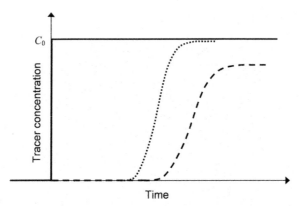

FIGURE 3.30   Breakthrough curves in a fracture subject to normal stress:
 solid line: injected particle concentration at the inlet;
 dotted line: small normal stress;
 dashed line: large normal stress.

Based on Koyama T, Li B, Jiang Y, Jing L. Numerical simulations for the effects of normal loading on particle transport in rock fractures during shear. Int J Rock Mech Min Sci 2008;45(8): 1403−19.

Particle retention decreases under these conditions since fewer particles get stuck in a wider fracture. This is schematically illustrated in Fig. 3.31, where two breakthrough curves are shown for two different shear displacement values.

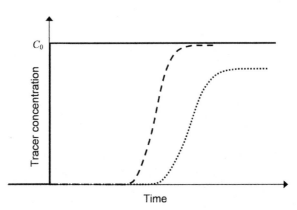

FIGURE 3.31   Breakthrough curves in a fracture undergoing dilatancy when subject to shear displacement:
 solid line: injected particle concentration at the inlet;
 dotted line: small shear displacement;
 dashed line: large shear displacement.

Based on Koyama T, Li B, Jiang Y, Jing L. Numerical simulations for the effects of normal loading on particle transport in rock fractures during shear. Int J Rock Mech Min Sci 2008;45(8): 1403−19.

Shear displacement brings about anisotropy in the flow patterns, discussed earlier in this chapter. As a result, particles travel faster in the direction normal to shear than in the direction parallel to shear. Fracture bridging and sealing with LCM particles will be affected by the direction of flow under such circumstances.

Surprisingly few studies have addressed transport of millimeter-size particles in rough-walled fractures. The sealing capacity of lost circulation materials is often tested in conduits having smooth (parallel or converging) walls. One exception is a thorough experimental study by Guo et al. [30]. In that study, different lost circulation materials were injected into rough-walled fractures produced in sandstone or shale blocks. The fractures were generated by injecting a fluid into a borehole drilled in the block. The fracture plane was thus parallel to the borehole axis. It was discovered that the fracture was sealed by particles predominantly in the area adjacent to the borehole. The quality of sealing was found to be nonuniform along the borehole. As a result, the weakest spot of the seal broke as the borehole pressure was increased. The fluid then started to flow preferentially through the broken spot, which eventually resulted in sealing of the spot by LCM particles again. This process repeated itself at different locations along the seal as the pressure rose. It was found that the sealing process was controlled by the particle concentration, the particle size distribution, and the composition of the LCM blend. It was also found that optimization of these parameters becomes more important for effectively sealing fractures with apertures greater than 1 mm. Fractures wider than 1 mm are more difficult to seal and require specialized LCM designs. Experiments on fracture bridging and sealing by LCM particles are further reviewed in chapters "Preventing Lost Circulation" and "Curing the Losses." Parameters measured in these experiments are the seal stability (the maximum pressure differential that the seal can withstand), the fluid loss as the seal builds up, and the time it takes to build the seal. Details of particle transport and deposition are rarely studied in the experiments performed with LCM. This is why, in the preceding discussion, we focused on the experiments performed with colloidal particles in environmental engineering.

In addition to the surface roughness, another factor affecting particle transport and deposition in fractures is the three-dimensional fracture shape and complex fracture networks. The Z-shaped fracture considered earlier in this chapter (Fig. 3.5) provides a good example. Particles transported in such fracture may come to a halt when they reach the kink if the aperture is smaller in the deviated part of the fracture than it is in the rest of the fracture. This situation, shown in Fig. 3.32, is believed to cause proppant screenouts in hydraulic fracturing [6]. In the context of LCM transport, such screenout might facilitate bridging of the fracture by LCM particles. Thus, three-dimensional fracture morphology plays an important role in lost circulation, affecting losses, on one hand, and the action of LCM, on the other.

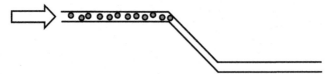

FIGURE 3.32   Particle transport in a Z-shaped fracture. The *arrow* indicates the flow direction.

3.5 SUMMARY

Natural fractures show a great variety in their morphology, roughness, permeability, connectivity, and deformability. This variety affects drilling in naturally fractured formations by making it difficult to predict the occurrence and severity of mud losses. Fracture permeability and a related property, hydraulic aperture, play a central role in lost circulation. Whether a natural fracture will cause losses depends on its aperture. Fracture aperture is affected by geological history (origin of the fracture, past shear and normal displacements, mineralization, etc.) and current in situ stresses. Once the drilling fluid has entered the fracture, its flow in the fracture is controlled by the hydraulic aperture and roughness of the fracture faces caused by asperities. Asperities make the fluid flow tortuous and reduce the fracture permeability. They also affect particle transport in the fracture (eg, transport and deposition of LCM particles). The variety of fracture properties makes it inherently difficult to predict how particles will behave once they have entered the fracture, for instance, whether they will build a seal and where such a seal will be built. This inherent complexity is exacerbated by poor fracture characterization. In particular, even though image logs may be used to pinpoint individual fractures and evaluate the fracture spacing, information about fracture apertures and roughness is rarely available. These factors contribute to inconsistent performance of lost circulation materials and other treatments. We will see in more detail in chapter "Mechanisms and Diagnostics of Lost Circulation" that natural fractures represent one of the most important mechanisms of lost circulation. Before proceeding to studying lost-circulation mechanisms, we briefly review functions and properties of drilling fluids in chapter "Drilling Fluid."

REFERENCES

[1] Twiss RJ, Moores EM. Structural geology. 2nd ed. W.H. Freeman; 2007. 500 p.
[2] Lespinasse M, Sausse J. Quantification of fluid flow: hydro-mechanical behaviour of different natural rough fractures. J Geochem Explor 2000;69−70:483−6.
[3] Schmittbuhl J, Steyer A, Jouniaux L, Toussaint R. Fracture morphology and viscous transport. Int J Rock Mech Min Sci 2008;45(3):422−30.
[4] MŁynarczuk M. Description and classification of rock surfaces by means of laser profilometry and mathematical morphology. Int J Rock Mech Min Sci 2010;47(1):138−49.

[5] Jeffrey RG, Bunger AP, Lecampion B, Zhang X, Chen ZR, van As A, et al. Measuring hydraulic fracture growth in naturally fractured rock. SPE paper 124919 presented at the 2009 SPE Annual Technical Conference and Exhibition held in New Orleans, Louisiana, USA, October 4–7, 2009.

[6] Daneshy AA. Off-balance growth: a new concept in hydraulic fracturing. J Petrol Technol 2003;55(4):78–85.

[7] Straub A, Krückel U, Gros Y. Borehole electrical imaging and structural analysis in a granitic environment. Geophys J Int 1991;106:635–46.

[8] Prensky SE. Advances in borehole imaging technology and applications. In: Lovell MA, Williamson G, Harvey PK, editors. Borehole imaging: applications and case histories, vol. 159. London: Geological Society; 1999. p. 1–43.

[9] Gutierrez M, Øino LE, Nygård R. Stress-dependent permeability of a de-mineralised fracture in shale. Mar Petrol Geol 2000;17(8):895–907.

[10] Koyama T, Li B, Jiang Y, Jing L. Numerical simulations for the effects of normal loading on particle transport in rock fractures during shear. Int J Rock Mech Min Sci 2008;45(8):1403–19.

[11] Esaki T, Du S, Mitani Y, Ikusada K, Jing L. Development of a shear-flow test apparatus and determination of coupled properties for a single rock joint. Int J Rock Mech Min Sci 1999;36(5):641–50.

[12] Chen Z, Narayan SP, Yang Z, Rahman SS. An experimental investigation of hydraulic behaviour of fractures and joints in granitic rock. Int J Rock Mech Min Sci 2000;37(7):1061–71.

[13] Barton N, Bandis S, Bakhtar K. Fundamentals of rock joint deformation. Int J Rock Mech Min Sci 1985;22(3):121–40.

[14] Zimmerman RW, Kumar S, Bodvarsson GS. Lubrication theory analysis of the permeability of rough-walled fractures. Int J Rock Mech Min Sci & Geomech Abstr 1991;28(4):325–31.

[15] Ozkan E, Raghavan R, Apaydin OG. Modeling of fluid transfer from shale matrix to fracture network. SPE 134830 paper presented at the SPE Annual Technical Conference and Exhibition held in Florence, Italy, September 19–22, 2010.

[16] Aybar U, Eshkalak MO, Sepehrnoori K, Patzek TW. The effect of natural fracture's closure on long-term gas production from unconventional resources. J Nat Gas Sci Eng 2014;21 :1205–13.

[17] Gangi AF. Variation of whole and fractured porous rock permeability with confining pressure. Int J Rock Mech Min Sci & Geomech Abstr 1978;15:249–57.

[18] Lavrov A. The Kaiser effect in rocks: principles and stress estimation techniques. Int J Rock Mech Min Sci 2003;40(2):151–71.

[19] Lavrov A. Fracture-induced physical phenomena and memory effects in rocks: a review. Strain 2005;41(4):135–49.

[20] Ameen MS. Fracture and in-situ stress patterns and impact on performance in the Khuff structural prospects, eastern offshore Saudi Arabia. Mar Petrol Geol 2014;50:166–84.

[21] Cuss RJ, Milodowski A, Harrington JF. Fracture transmissivity as a function of normal and shear stress: first results in Opalinus Clay. Phys Chem Earth Parts A/B/C 2011;36 (17–18):1960–71.

[22] Frigaard IA, Ryan DP. Flow of a visco-plastic fluid in a channel of slowly varying width. J Non-Newtonian Fluid Mech 2004;123(1):67–83.

[23] Lavrov A. Non-Newtonian fluid flow in rough-walled fractures: a review. In: Feng X-T, Hudson JA, Tan F, editors. The 3rd ISRM Symposium on rock mechanics SINOROCK 2013. Taylor & Francis; 2013. p. 363–8.

[24] Mitsoulis E, Huilgol RR. Entry flows of Bingham plastics in expansions. J Non-Newtonian Fluid Mech 2004;122:45−54.

[25] Lavrov A. Numerical modeling of steady-state flow of a non-Newtonian power-law fluid in a rough-walled fracture. Comput Geotech 2013;50:101−9.

[26] Lavrov A. Redirection and channelization of power-law fluid flow in a rough-walled fracture. Chem Eng Sc 2013;99:81−8.

[27] Cumbie DH, McKay LD. Influence of diameter on particle transport in a fractured saprolite. J Contam Hydrol 1999;37:139−57.

[28] Alaskar M, Li K, Horne R. Influence of particle size on its transport in discrete fractures: pore-scale visualization using micromodels. Paper SGP-TR-198 presented at the Thirty-Eighth Workshop on Geothermal Reservoir Engineering held at Stanford University, Stanford, California, February 11−13, 2013.

[29] Bauget F, Fourar M. Non-Fickian dispersion in a single fracture. J Contam Hydrol 2008;100:137−48.

[30] Guo Q, Cook J, Way P, Ji L, Friedheim JE. A comprehensive experimental study on wellbore strengthening. IADC/SPE paper 167957 presented at the 2014 IADC/SPE Drilling Conference and Exhibition held in Fort Worth, Texas, USA, March 4−6, 2014.

Chapter 4

Drilling Fluid

Lost circulation means loss of drilling fluid into the formation. Therefore, before proceeding to the discussion of lost circulation and its mechanisms, the basic functions, properties, composition, and behavior of drilling fluids are first reviewed in this chapter. Emphasis is placed on those properties that are most important for lost circulation, such as the density and rheological properties (yield stress, plastic viscosity, thixotropic properties). The chapter is not a replacement for a standard text on drilling fluid composition and properties, and should be considered only a summary of the most basic facts without which the subsequent chapters cannot be understood. The reader is encouraged to consult standard texts on drilling fluid engineering, eg [1,2], for an in-depth treatment of the material covered in this chapter.

4.1 FUNCTIONS OF DRILLING FLUID

Drilling fluid has the following basic functions:

- prevention of formation fluid influx into the well;
- application of stress on the borehole wall in order to prevent borehole instability (breakouts, collapse);
- lubrication and cooling of the drill string;
- cleaning of the borehole and prevention of cuttings accumulation at the bit;
- cuttings transport out of the well.

In addition, when drilling through a producing zone, the drilling fluid composition should be such as to minimize formation damage. Since some of the above goals call for opposite changes in drilling fluid properties and composition, designing a drilling fluid for a specific formation is an optimization exercise. This partly explains why a great variety of drilling fluids are available on the market, and why dozens of different ingredients are used in various proportions in these products.

4.2 PROPERTIES OF DRILLING FLUIDS

The main properties of a drilling fluid are [2]:

- density (mud weight)
- rheological properties (plastic viscosity, yield stress, gel strength)

Lost Circulation. http://dx.doi.org/10.1016/B978-0-12-803916-8.00004-2

81

- fluid loss properties
- stability of the drilling fluid with respect to elevated temperature and pressure
- stability of the drilling fluid with respect to contaminating fluids, such as salt water.

In overbalanced drilling, the density of the drilling fluid must be sufficiently high to avoid borehole instabilities and to prevent formation fluid influx. At the same time, the density of the drilling fluid should usually be kept below a certain threshold (the *lost-circulation pressure*) in order to prevent mud losses, or at least keep them under control. The density of the drilling fluid can be adjusted by adding barite or other weighting materials such as calcium carbonate or hematite.

Drilling fluids usually have yield-stress rheology. The simplest yield-stress rheological model is the Bingham fluid described in chapter "Natural Fractures in Rocks" and widely used in drilling engineering. The Bingham model has only two parameters, the yield stress and the plastic viscosity (Fig. 4.1). The yield stress describes the flow-resisting force that can be caused, for example, by electrochemical interaction between clay particles in a water-base mud. The plastic viscosity describes the frictional forces when the fluid flows. Both parameters can be obtained from viscometric measurements. In particular, the yield stress is estimated by performing measurements at two rpm values and then extrapolating the shear stress so as to estimate the value it would have at zero shear rate. The use of this technique (and of viscometers in general) for yield stress evaluation is a debated issue since it can provide only a rough approximation. The accuracy can be improved by using more sophisticated testing devices (rheometers). However, this improvement does not eliminate the fundamental shortcoming, namely the use of a *dynamic* technique to estimate an essentially *static* parameter (the yield stress).

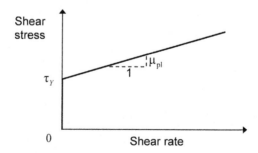

FIGURE 4.1   Shear stress versus shear rate for a Bingham fluid. Yield stress and plastic viscosity are shown by τ_Y and $|\dot{\gamma}|$, respectively.

A radically different approach to measuring the yield stress was proposed by Majidi et al. [3]. If a yield-stress fluid is injected from a point source into a slot of constant aperture, it will flow in a radial pattern. The flow will be divergent (the front will be expanding). The thickness of the solid plug of unyielded material (Fig. 3.26) will therefore increase with distance from the source. Eventually, it will expand to such extent that the unyielded material will occupy the entire aperture of the slot. At this point, the flow will stop since the overpressure at the source is balanced by the yield stress of the fluid occupying the slot. Measuring the final radius of the invaded zone, the yield stress of the fluid can be back calculated. This arguably provides a more accurate estimate of the yield stress than the dynamic measurements in a viscometer. Moreover, flow in a slot is quite similar to flow in a fracture, although a real fracture would have rough walls and varying aperture. Investigating mud flow into a slot under laboratory conditions provides an insight into what happens when a natural fracture is met during overbalanced drilling.

The Bingham model is most commonly used to describe the drilling fluid rheology in the industry, but it is known to yield a poor description of the fluid rheology at low shear rates [4]. A slightly more sophisticated and often more accurate model is the *Herschel–Bulkley model*. In a simple shear flow, the shear stress depends on the shear rate as follows in the Herschel–Bulkley fluid:

$$\tau = \tau_Y + K|\dot{\gamma}|^n \qquad [4.1]$$

where τ is the shear stress; $\dot{\gamma}$ is the shear rate; K is the consistency index; n is the power-law index; and τ_Y is the yield stress.

The Bingham model can be obtained from the Herschel–Bulkley model by setting $n = 1$. Another specific case of the Herschel–Bulkley model is obtained by setting $\tau_Y = 0$. This fluid is called (simple) *power-law fluid*, or Ostwald–de Waele fluid. This model has two parameters, K and n. A specific case of Ostwald–de Waele fluid is *Newtonian fluid*, obtained with $n = 1$. In the case of a Bingham fluid, K is the plastic viscosity. In the case of a Newtonian fluid, K is called the *dynamic viscosity* of the fluid.

A drilling fluid needs to have a certain viscosity and a certain yield stress to be able to transport drill cuttings out of the hole. This is usually achieved by adding viscosity control agents; eg, clays such as bentonite, attapulgite, or sepiolite [5]. In a sufficiently permeable rock, clay particles build a filter cake on the borehole wall preventing fluid loss and pressure diffusion into the rock. Moreover, clay content and type determine the gel strength and the yield stress of the fluid. The gel strength and yield stress must be kept above a certain threshold to ensure that drill cuttings and barite particles are held in suspension when circulation stops.

Time-dependent behavior of drilling fluids is described by their *gel strength*. The gel strength builds with time as a result of interaction between clay particles suspended in the fluid. This buildup is a consequence of *thixotropic behavior* of the drilling fluid, which makes its rheological properties depend on the duration of shearing. When the fluid is at rest, the interaction between clay particles leads to the development of internal structures (*flocculation*) which increases the resistance of the fluid to the initiation of shearing. The gel strength describes the strength of such interparticle interaction under static conditions. The distinction is usually made between the *initial gel* measured after the fluid has been static for 10 s, and the *10-min gel* measured after the fluid has been static for 10 min. Drilling fluids with a large difference between the 10-min and 10-s gel are said to have "*progressive gel.*" In practice, the value of gel strength can be obtained from viscometric measurements of the shear stress at low shear rates after a period of quiescence. The gel strength prevents settling of suspended solids when circulation is stopped. Buildup of the gel strength with time may create problems when resuming the circulation if the bottomhole pressure increases above the lost-circulation pressure. The gel strength can be reduced; eg, by adding *deflocculants.* This reduces the flocculation capacity of fines. The gel strength can be also reduced by diluting the drilling fluid. This reduces the solids content.

Gel strength is a common engineering way to describe the thixotropic behavior of drilling fluids. Thixotropy means, in particular, that if a constant shear rate is applied over some time, the shear stress will gradually decrease, reaching an asymptotic value typical of that shear rate. A consequence of thixotropy is a hysteresis observed when the shear rate is periodically ramped up and down in a rheometric experiment [4].

The rheological properties of drilling fluids are affected by pressure and temperature. The temperature and pressure dependence of the fluid properties are important for drilling of high-pressure, high-temperature (HPHT) wells. Generally, the yield stress and the plastic viscosity of a drilling fluid decrease with temperature and increase with pressure. The *failure temperature* can be defined as the temperature at which the drilling fluid changes its rheological properties to the extent that prevents its further use [6]. Typically, oil-base muds have higher failure temperatures than water-base muds, under the same pressure.

4.3 COMPOSITION OF DRILLING FLUIDS

In this section, we will see how properties of the drilling fluids described in Section 4.2 are affected by the fluid composition. The section provides the basic information about drilling fluid composition required to understand subsequent chapters. In-depth coverage of drilling fluid composition and chemistry can be found, for example, in Refs [1,2,5].

A classification of drilling fluids with regard to their composition is shown in Fig. 4.2. All-liquid drilling fluids can be broadly classified according to their base fluid (the continuous phase) into *water-base muds (WBMs), oil-base muds*

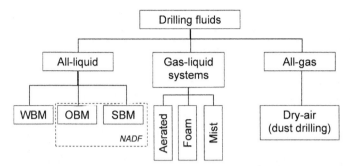

FIGURE 4.2 Classification of drilling fluids according to their composition. *WBM*, water-base mud; *OBM*, oil-base mud; *SBM*, synthetic-base mud; *NADF*, nonaqueous drilling fluid.

(OBMs), and *synthetic-base muds (SBMs)*. The latter two are collectively known as *nonaqueous drilling fluids (NADFs)*.

The base fluid in WBM is fresh water, seawater, or saltwater. WBM may contain oil in dispersed form. WBMs can be divided into polymer-based fluids and clay-based fluids, depending on the agents used to build the rheology of the fluid. The structure built in the fluid by charged clay particles (or by polymer molecules) increases the low-rate viscosity and creates the yield stress. As the shear rate increases, the structure breaks and the fluid shear-thins.

The base fluid in an OBM can be crude oil, diesel oil, or mineral oil. An OBM may also contain water in dispersed form. If the water content in an OBM exceeds 5%, the mud is classified as an invert emulsion mud (water in oil).

A number of compounds have been proposed as the base fluids in SBMs, in particular synthetic hydrocarbons.

In addition to the three types of all-liquid fluids shown in Fig. 4.2, reversible drilling fluids have been developed. The continuous phase in a reversible fluid can be switched from water to oil (or synthetic), or from oil (or synthetic) to water.

Typical ingredients of a WBM include the following [2]:

- fresh water, seawater, or saltwater as the base fluid;
- viscosity control agents;
- weighting agents;
- fluid loss control agents;
- emulsifiers;
- lubricants;
- corrosion inhibitors;
- salts and alkalies;
- pH control agents.

Typical ingredients of an OBM include the following [2]:

- oil as the base fluid;
- dispersed water with dissolved salts and alkalies;
- viscosity control agents;
- weighting agents;
- fluid loss agents;
- emulsifiers;
- oil-wetting agents.

Typical ingredients of gas–liquid systems include the following:

- gas;
- water;
- surfactants;
- polymers;
- possibly salts.

Compared with WBMs, OBMs are superior in terms of lubrication properties, shale inhibition, rate of penetration, reduced formation damage, reduced differential sticking, reduced corrosion, and enhanced thermal stability. They also have better resistance to contaminants such as salts, H_2S, and CO_2 [6]. Their use, however, is restricted by environmental regulations.

Water-base muds are typically considered to be more environmentally friendly. However, their use in HPHT wells requires that their thermal stability be improved, and this might necessitate using additives that jeopardize their low environmental impact [6].

OBMs, in addition to their environmental impact, may have a detrimental effect on rubber parts of equipment. They also affect the performance of formation logging tools that make use of conductive drilling fluid in the well. Barite sagging is more severe in OBM than in WBM [4]. Nonaqueous drilling fluids (OBM and SBM) are known to be considerably more compressible than WBM. Thermal stability of nonaqueous fluids is typically better than that of WBMs, but their properties, in particular the viscosity, are usually strongly affected by the temperature and pressure. Thermal stability of nonaqueous drilling fluids makes them a preferred choice in HPHT wells. However, detection of gas kicks is more difficult with nonaqueous drilling fluids than with WBM because of gas solubility in NADF. As a result, well control becomes more difficult with NADFs [7]. Moreover, the initial costs of NADFs are higher than those of WBMs. NADFs are therefore typically used in rock formations where the increased mud costs and cuttings disposal costs are offset by the benefits of NADFs. Examples are shale or salt formations, high-temperature formations, production zones where formation damage must be minimized, etc. Many of the shortcomings of nonaqueous drilling fluids mentioned above can be offset by specially designed additives [8]. This, however, is likely to increase the cost of the fluid.

Various additives are used to improve the performance of WBMs bringing it to the level of NADFs. Thermal stability of WBMs can be improved by adding formates (potassium formate or sodium formate) as well as polysaccharides. To improve shale-inhibiting properties of water-base muds, salts of potassium, such as potassium chloride and potassium acetate, are used as shale inhibitors. These salts also reduce dispersion of shale particles in the mud. Alternatively, other shale inhibitors can be used in WBM; eg, polyols and sodium or potassium silicates. The latter create a water barrier on shale upon gelation, thereby preventing hydration of the shale. This can reportedly provide shale inhibition capacity compared with that of OBM [2]. NaCl and $CaCl_2$ are added also to the water phase of oil-base muds to prevent swelling of active shales.

Emulsifiers are surfactants used to prevent coalescence of emulsions. Coalescence is enhanced at elevated temperatures and can alter rheological properties of the fluid. Calcium and magnesium fatty acid soaps are commonly used as emulsifiers in oil-base muds.

Another type of surfactant used in oil-base muds is wettability reversal agents. Mineral particles introduced into the mud as the drill bit breaks the rock are usually water-wet. As a result, these solids tend to be agglomerated by water droplets in the water-in-oil emulsion. Such agglomeration would normally increase the mud viscosity and promote barite settling. To prevent this from happening, wettability of the solids is changed from water-wet to oil-wet by adding wettability reversal additives to the oil phase of the OBM.

Clays that readily hydrate, in particular sodium montmorillonite, are widely used in the industry to increase viscosity, yield stress, and the cuttings carrying capacity of water-base muds. Bentonite is a commercially available product composed primarily of sodium montmorillonite. Other viscosity control agents are polymers (xanthan gum, polyacrylamide, polyurethanes, polyesters, carboxymethylcellulose, hydroxyethylcellulose, and various kinds of starch).

In oil-base muds, asphalts and amine-treated bentonite are used as viscosity control agents. Viscosity of oil-base fluids is also increased by water droplets present in the emulsion, and by soaps used as emulsifiers.

Phosphates, lignite, and lignosulfonates are frequently used in water-base muds as thinners due to their deflocculating action on suspended clay solids. Deflocculation reduces the gel strength and the yield stress of the mud.

Filtration control additives (fluid loss control agents) are frequently used in water-base muds to minimize the loss of the base fluid (water) into the formation. Fluid loss control agents used in water-base muds include different kinds of polymers such as starch, polyanionic cellulose, carboxymethylcellulose, and sodium polyacrylate. The reduction in fluid loss is achieved by polymers increasing the viscosity of the base fluid (water), thereby reducing the filtration rate through the borehole wall. Performance of the polymers deteriorates at high temperature and salinity.

Bentonite and deflocculants such as lignites reduce the fluid loss as well. The advantages of lignites are their temperature stability and thinning properties. Lignites improve the fluid loss properties of the mud without raising its viscosity.

Oil-base fluids have superior fluid loss properties and can be deployed without adding fluid loss control agents. Bentonite and water droplets present in OBM can build a high-quality, low-permeability, thin filter cake on the borehole wall. Excessive overbalance may, however, squeeze the water droplets into the porous rock, with detrimental effects on the permeability of the pay zone. In situations where the fluid-loss properties of OBM need to be improved, asphalts, polymers, amine-treated lignite, and other additives can be used.

Density control is accomplished by adding weighting agents to the mud. The most commonly used such agent, both in oil- and water-base muds, is barite (barium sulfate), an inert material with the specific gravity of 4.2 and the particle size within the range of approximately $1-100$ μm, with the median size $15-20$ μm. A side effect of adding barite is an increase of the mud viscosity. Additional complications of using barite in oil-base fluids may be barite settling due to low gel strength of OBM, and barite agglomeration due to its incomplete wettability reversion to oil-wet. Instead of barite, heavier and harder density-control agents, such as minerals hematite and ilmenite, have been used in water-base fluids. Moreover, microfine barite having the median particle size of 1.8 μm has been introduced [4].

Drilling fluids may have pH in the range of c.7 to 13 to prevent corrosion and hydrogen embrittlement. Alkalinity control is achieved by adding caustic to water-base muds or lime to oil-base muds.

In addition to the original composition, the mud returning from the borehole contains formation fluids and drill cuttings. Cuttings such as sand or feldspar particles have detrimental effect on the mud. In particular, they increase the frictional pressure losses and reduce the quality of the filter cake by making it thicker and more permeable. Undesired solids can be removed from the mud by means of vibrating screens, hydrocyclones, centrifuges, and chemical treatments.

The environmental impact of OBMs mentioned earlier can be reduced by using synthetic-base muds. These muds have most of the advantages of traditional OBMs. As the base fluid, synthetic organic fluids—eg, derived from ethylene—are used in synthetic-base muds. Synthetic-base muds are superior with regard to their performance, compared with WBMs. Their disadvantage is high initial cost. Some synthetic muds also exhibit reduced stability (ester-based muds). The additives used in synthetic muds are similar to those used in OBMs; ie, viscosifiers, emulsifiers, and weighting agents.

In addition to fluids with both the base fluid and the dispersed fluid being liquids, gas—liquid systems and all-air systems are in use (Fig. 4.2). Underbalanced drilling enabled with such fluids reduces the risk of lost circulation. Some applications of such fluids are described in chapter "Preventing Lost Circulation."

4.4 SOLID PARTICLES: SIZE, SHAPE, AND SETTLING

As we have seen in Section 4.3, drilling fluids contain solid particles. For instance, bentonite and treated bentonite are used to build a foundation of rheology, preventing settling of barite in water- and oil-base muds, respectively. Weighting agents such as barite are another example of solid particles. Materials used to cure lost circulation are often particulates or fibers. Drill cuttings are particles of different shapes and sizes.

The shape and size of particles in a drilling fluid differ widely and need to be quantified because they eventually determine the properties of the drilling fluid and the behavior of the particles.

There are several ways to describe the size of an individual solid particle of, in general, an irregular shape [9]:

- the sieve size;
- the equivalent diameter determined from the particle volume or surface;
- the equivalent diameter determined from the settling velocity of the particle.

The sieve size is the width of the minimum square aperture through which the particle can pass. The equivalent size based on the particle volume is the diameter of a spherical particle that has the same volume as the irregularly shaped particle; ie,

$$D_v = (6V_p/\pi)^{1/3} \qquad [4.2]$$

where D_v and V_p are the equivalent diameter and the volume of the irregularly shaped particle, respectively. The equivalent size based on the particle surface is the diameter of a spherical particle that has the same surface area as the irregularly shaped particle; ie,

$$D_s = (S_p/\pi)^{1/2} \qquad [4.3]$$

where D_s and S_p are the equivalent diameter and the surface area of the irregularly shaped particle, respectively.

The equivalent size based on the settling velocity is the diameter of a spherical particle that has the same terminal settling velocity in a viscous fluid as the irregularly shaped particle.

In reality, particles of very different sizes are present in a drilling fluid. *Particle size distribution (PSD)* can be used to fully describe the size of the particles. The PSD shows the mass (or volume) fraction of particles with the diameter smaller than the given value. The median size—ie, the particle size such that the fraction of particles smaller than that is 50%—is usually denoted D_{50}. Specifying the value of D_{50} is the simplest way to describe the particle size distribution with a single number. For instance, barite particles typically are within the range from 1 to 100 μm [5]. The value of D_{50} is around

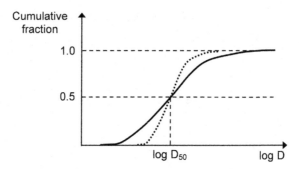

FIGURE 4.3 Relatively broad (*solid line*) and relatively narrow (*dotted line*) particle size distributions.

$10-20\ \mu m$ for barite. The value of D_{50} provides, however, no information about how wide or how narrow the PSD is. As an illustration, two particle size distributions are shown in Fig. 4.3. The two distributions have the same value of D_{50}, but the one represented by the dotted line is narrower (more uniform; ie, better sorted) than the one represented by the solid line.

In addition to the particle size, the *particle shape* is of significance. The particle shape can be quantified by *sphericity*, Ψ, defined as [9]:

$$\Psi = D_v^2/D_s^2 = \pi(6V_p/\pi)^{2/3}/S_p \qquad [4.4]$$

For a sphere, $\Psi = 1$. For angular particles, $0 < \Psi < 1$. For instance, for a particle shaped as a cube: $\Psi = (\pi/6)^{1/3} \approx 0.806$. Particle sphericity may affect settling by altering the terminal settling velocity and making the particle assume a preferential orientation during settling.

The particle size distribution and the particle shape have a major impact on particle deposition and packing. Close packing of equal spheres has a solids volume fraction of 0.74. The *solids volume fraction* is defined as the ratio of the volume occupied by particles to the total volume. Random packing of equal spheres may produce solids volume fractions from 0.59 to 0.64 [9]. In the vicinity of a wall the packing is disturbed. The solids volume fraction drops below the theoretical value near the wall. As a result, the solids volume fraction is higher in the bulk than near a wall. This is a pure geometry effect well-known in granular systems.

The solids volume fraction decreases with decreasing sphericity and increasing roughness of the particles.

In the presence of particles of different sizes, the solids volume fraction may increase above the theoretical value derived for equal spheres. In this case, the smaller particles occupy part of the void space left in between the larger particles, thereby reducing the porosity and increasing the solids volume fraction.

Particle behavior in a solid—liquid system such as drilling mud is governed by a number of forces. An in-depth coverage of these forces can be found in Refs [9,10]. Here, we consider only one aspect of particle motion in a fluid, namely particle settling in a quiescent fluid. An oilfield example of this motion would be settling of solids in the well after the circulation is stopped. Consider first settling of a single spherical particle in a Newtonian fluid. The velocity of the particle settling in a quiescent fluid will asymptotically approach the *terminal settling velocity*. The latter is determined by the balance of the particle gravity, the particle buoyancy, and the viscous drag force. The magnitude of the viscous drag force, F_D, is given by [9].

$$F_D = \frac{1}{2} C_D A_p \rho_f v_r^2 \tag{4.5}$$

where C_D is the drag coefficient (dimensionless); v_r is the magnitude of the particle velocity relative to the fluid; A_p is the area of the particle projection on the plane normal to the relative velocity vector, $\mathbf{v_r}$; ρ_f is the density of the fluid. For a particle moving with the terminal settling velocity, v_∞, this velocity can be obtained by equating the drag force given by Eq. [4.5], to the sum of the gravity and buoyancy forces. This yields:

$$v_\infty = \left[\frac{2(\rho_s - \rho_f)g V_p}{C_D A_p \rho_f} \right]^{1/2} \tag{4.6}$$

where ρ_s is the density of the particle; g is the acceleration of gravity. For a spherical particle of diameter D, Eq. [4.6] becomes

$$v_\infty = \left[\frac{4(\rho_s - \rho_f)g D}{3 C_D \rho_f} \right]^{1/2} \tag{4.7}$$

The drag coefficient in Eqs. [4.5]—[4.7] is a function of the particle shape, the proximity to walls, the presence of other particles (hindered settling), the flow regime (turbulent vs. laminar), and the fluid rheology. Even in the simplest case of a single spherical particle having the diameter much smaller than the distance to walls, numerous expressions for the drag coefficient as a function of particle's Reynolds number have been proposed. One classical equation for the particle drag coefficient in a Newtonian fluid is given by [9].

$$C_D = \frac{24}{Re_p} \qquad\qquad \text{for } Re_p < 0.2$$

$$C_D = \frac{24}{Re_p}\left(1 + 0.15 Re_p^{0.687}\right) \quad \text{for } 0.2 < Re_p < 1000 \tag{4.8}$$

$$C_D = 0.44 \qquad\qquad \text{for } 1000 < Re_p < 3 \cdot 10^5$$

with the particle's Reynolds number defined as

$$Re_p = \frac{v_r \rho_f D}{\mu} \qquad [4.9]$$

where μ is the dynamic viscosity of the fluid.

The effect of particle shape on C_D is schematically illustrated in Fig. 4.4. Irregularly shaped particles exhibit higher drag coefficient than rounded particles at the same Re_p.

A survey of equations for C_D for a Newtonian fluid can be found in Refs [9] and [11]. In non-Newtonian fluids, the issue becomes significantly more complicated. A commonly used, engineering approach makes use of the same functional dependency of C_D on Re_p as for a Newtonian fluid, but with the particle Reynolds number re-defined so as to account for the fluid rheology. To this end, the dynamic viscosity, μ, in Eq. [4.9] should now be thought of as the apparent viscosity of the non-Newtonian fluid, ie, the ratio of the shear stress to the shear rate. For the shear rate, the ratio of the particle relative velocity to its size is usually used; ie, $|\dot{\gamma}| = v_r/D$. For a Bingham fluid, this approach results in the following expression for the particle Reynolds number [11]:

$$Re_p = \frac{v_r \rho_f D}{\mu_{pl} + \tau_Y D/v_r} \qquad [4.10]$$

An example of an empirical C_D versus Re_p dependency that fits both Newtonian and at least some non-Newtonian fluids is given by [11].

$$C_D = \frac{24}{Re_p}\left(1 + a_1 Re_p^{a_2}\right) + \frac{a_3}{1 + a_4/Re_p} \quad \text{for} \quad 0.1 < Re_p < 1000 \qquad [4.11]$$

where a_1, a_2, a_3, a_4 are nondimensional fitting parameters. The values of $a_1 = 0.1407$, $a_2 = 0.6018$, $a_3 = 0.2118$, $a_4 = 0.4215$ provide a reasonably accurate approximation to the experimental data for a Newtonian fluid and some non-Newtonian shear-thinning fluids without yield stress [11]. The

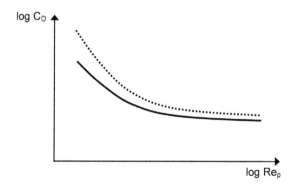

FIGURE 4.4 Schematic illustration of the drag coefficient versus the particle's Reynolds number for a more rounded particle (*solid line*) and a more angular particle (*dotted line*).

validity of Eq. [4.11] for yield-stress fluids, such as drilling muds, remains to be established.

Additional complications arise in yield-stress fluids when several particles settle one after another along the same trajectory. Experiments performed with solid spheres in polymer dispersions (Carbopol) have shown that the terminal settling velocity of each subsequent particle is higher than that of the preceding particle, until it eventually levels off at some asymptotic value. In the experiments reported in Ref. [12], this asymptotic value was reached after three to four particles. A rest period follows, the fluid shows an ability to "heal": if the experiment is repeated after the rest period, the first particle again has a lower terminal settling velocity. The terminal velocity of subsequent particles then again increases toward the asymptotic value. These results suggest that the structure of the polymer solution is disturbed by settling particles, which affects the apparent viscosity experienced by each subsequent particle. A period of rest applied between particle releases enables the fluid's structure to "heal," thereby restoring the terminal settling velocity to its original value typical of a "fresh" fluid.

Particle settling in a confined environment—eg, in a fluid-filled rock fracture—will be affected by the proximity to walls. The walls reduce the settling velocity in a quiescent Newtonian fluid. In the case of a quiescent yield-stress fluid, the shear stresses exceed the yield stress of the fluid only within a shear zone near the settling particle. Outside the shear zone, there is no flow; ie, $|\dot{\gamma}| = 0$. Therefore, there is no effect of the wall unless the particle approaches the wall so closely that the wall happens to be within the shear zone [12].

An important feature of settling in a yield-stress fluid is that settling only occurs with particles greater than a certain critical size. This is remarkably different from Newtonian fluids. Settling requires that shear is induced around the particle. In a yield-stress fluid, it implies that the shear stresses exceed the yield stress of the fluid, τ_Y. For a sufficiently small particle, the shear stress induced by gravity (minus buoyancy) is not sufficient to generate shear stresses in excess of τ_Y. Thus, such particles cannot settle and will stay in suspension indefinitely.

A simple estimation of the critical size, D_{min}, that a spherical particle needs to have in order to settle in a yield-stress fluid has been proposed in Ref. [9]. Assume that, for a particle of critical size, shear is initiated simultaneously over the entire particle surface. The critical particle size is obtained by equating the integral of the shear stress component acting on the particle upwards, to the particle's gravity (minus buoyancy) [9]:

$$\frac{1}{2} \pi D_{min}^2 \tau_Y \int_{-\pi/2}^{\pi/2} \cos^2\theta \, d\theta = \frac{1}{6} \pi D_{min}^3 (\rho_s - \rho_f) \qquad [4.12]$$

where the sense of the integration angle, θ, is indicated in Fig. 4.5.

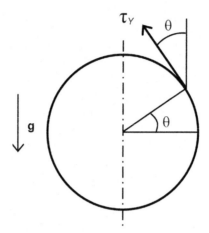

FIGURE 4.5   Illustration to the derivation of the critical particle size for settling in a yield-stress fluid.

The critical particle size is thus given by

$$D_{min} = \frac{3\pi\tau_Y}{2(\rho_s - \rho_f)g}$$ [4.13]

Eq. [4.13] is only a crude approximation. More accurate finite-element estimations of the flow field around a spherical particle settling in a Bingham fluid suggest that the critical particle diameter is given by [13].

$$D_c = \frac{3\tau_Y}{0.143(\rho_s - \rho_f)g}$$ [4.14]

For a Newtonian fluid ($\tau_Y = 0$), the critical particle size is zero, as expected; ie, particles of any size will settle. For a yield-stress fluid, the critical size increases with τ_Y. Thus, the thicker the fluid, the larger particles it can hold in suspension. This is the reason why drilling fluids are designed to have yield-stress rheology: the yield stress keeps solids in suspension when the circulation is stopped. Eq. [4.14] also indicates that, for a particle of a given size, D, a critical yield stress value exists given by

$$\tau_{Yc} = 0.04767(\rho_s - \rho_f)gD$$ [4.15]

If the yield stress of the fluid exceeds the critical value, the particle will not settle. The drag coefficient, C_D, becomes infinite in this case.

If the particle size exceeds the critical value given by Eq. [4.14], settling takes place. For a spherical particle moving with the terminal settling velocity in a yield-stress fluid, the flow field around the particle consists of three distinct regions, as illustrated in Fig. 4.6. Unyielded regions ("polar caps") exist at the stagnation points where the shear stress cannot overcome the yield

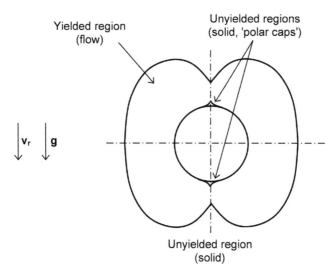

FIGURE 4.6 Yielded and unyielded regions around a spherical particle settling in a quiescent yield-stress fluid.

stress of the fluid. An unyielded region also exists in all the fluid outside the outer yield boundary. The Bingham material behaves like a solid outside the outer yield boundary. Between that boundary and the particle, the Bingham material is yielded and behaves like a liquid. For a given particle, as the yield stress of the fluid increases, the yielded region in Fig. 4.6 shrinks until it vanishes when τ_Y becomes equal τ_{Yc} given by Eq. [4.15].

4.5 HOW DOES DRILLING FLUID WORK?

The rheological behavior of drilling fluids is typically shear-thinning; ie, the apparent viscosity decreases with the shear rate. The apparent viscosity is the ratio of the shear stress to the shear rate, at a given value of the latter. The apparent viscosity can be illustrated graphically as the slope of a straight line joining the origin with the point on the shear stress versus shear rate curve corresponding to the shear rate of interest (Fig. 4.7). Thus, the apparent viscosity of a non-Newtonian fluid is the viscosity of a Newtonian fluid that would exhibit the same shear stress at the given shear rate.

Shear-thinning behavior of drilling fluids has practical implications. For instance, in direct circulation, when the drilling fluid flows downwards inside the drill string, its velocity is relatively high, and the apparent viscosity is relatively low. When the fluid returns up the annulus, its velocity is lower, and the apparent viscosity is larger. This helps the fluid transport the drill cuttings picked at the drill bit.

With regard to mud losses, the shear-thinning behavior of drilling fluids explains why, sometimes, it is possible to have only minor losses while drilling

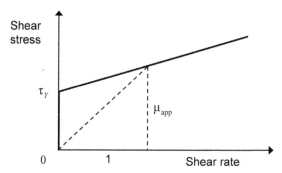

FIGURE 4.7 Definition of apparent viscosity, μ_{app}, for a yield-stress fluid. The yield stress is shown by τ_Y.

with overbalance in naturally fractured formations. When the borehole hits a natural fracture, the fluid starts flowing from the hole into the fracture. The drilling fluid starts displacing the formation fluid. The fluid displacement front spreads and becomes larger as more mud is lost into the fracture. The expansion of the fluid displacement front is due to the approximately radial geometry of the flow from the borehole into the fracture. Such flow can be thought of as radial flow from a point source located in the fracture plane. Eventually, the pressure gradient at the fluid displacement front will drop below the critical value given by $2\tau_Y/w$, where τ_Y is the yield stress of the drilling fluid; w is the fracture aperture at the location of the fluid displacement front. Consequently, losses will stop [8]. A certain finite amount of the drilling fluid will be lost into the fracture, creating formation damage. The incident will, however, not cause a total loss of the mud column. The loss will stop in this case even if no filter cake builds in the fracture, and no lost circulation material is used in the drilling fluid. The mud loss stops solely because of the yield-stress rheology of the drilling fluid. The loss would stop even if the fracture had perfectly impermeable walls, as long as the fluid has a nonzero yield stress and the overbalance is not high enough to increase the fracture aperture, or create new fractures, or propagate the existing fractures. However, if the aperture of the fracture is significantly increased by the overbalance (ie, the fracture opens wider), the threshold pressure gradient given by $2\tau_Y/w$ will decrease, and the loss might not stop. This normally happens when the wellbore pressure exceeds the fracture reopening pressure (chapter: Mechanisms and Diagnostics of lost Circulation).

Transport of drill cuttings out of the hole is one of the primary functions of the drilling fluid. The settling properties of solid particles carried by the fluid are central to the ability of the fluid to effectively clean the hole. Consider an illustrative example discussed by Tehrani [4]: When drilling a horizontal section of length L, the fluid must be able to carry particles out of the horizontal section before they settle down. The settling time is on the order of

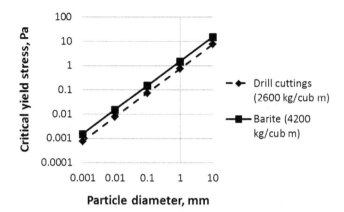

FIGURE 4.8 Critical yield stress required to hold solid spherical particles in suspension, based on Eq. [4.15].

$t_s = D_w/v_\infty$, where D_w is the wellbore diameter. The time needed to transport the particle along L is given by $t_t = L/v_a$ where v_a is the annular fluid velocity. Effective hole cleaning requires that the settling time be much larger than the transport time; ie,

$$v_\infty << \frac{D_w}{L} v_a \qquad [4.16]$$

The rheology of the drilling fluid should be sufficiently high in order to reduce the settling velocity below the limit given by the right-hand side of Eq. [4.16]. As pointed out by Tehrani [4], the rheology alone might not be sufficient to enforce the condition given by Eq. [4.16]. Pipe rotation and an increase in the pumping rate can be used to ensure good hole cleaning, in addition to increasing the fluid rheology.

Yield-stress rheology of drilling fluids plays a central role in preventing solid particles from settling when circulation is stopped. As discussed in Section 4.4, a spherical particle does not settle when the yield stress of the carrier fluid is greater than the critical value, τ_{Yc}. The dependence of this critical yield stress on the particle size, given by Eq. [4.15], is illustrated in Fig. 4.8 for spherical particles of two densities: 2600 kg/m³, typical of drill cuttings, and 4200 kg/m³, typical of barite. Suspending 10-mm cuttings requires the yield stress on the order of 10 Pa, according to Fig. 4.8 and Eq. [4.15]. Suspending 100-μm barite requires the yield stress of 0.1 Pa.

4.6 SUMMARY

The brief overview of drilling fluids presented in this chapter shows the enormous variety in their composition and properties. This variety contributes to the complexity of lost circulation problems. It gives rise to different techniques used to prevent and cure losses. For instance, varying base fluids enables control of mud weight within a wide range, including underbalanced drilling with gas—liquid muds. Adjusting rheological properties of the drilling fluid can be used to reduce the bottomhole circulating pressure and thus reduce or stop mud losses, etc. In the next three chapters, we will see how the properties and behavior of drilling fluids affect lost circulation and how they can be used to prevent or cure this drilling problem. We start with a detailed discussion of lost circulation mechanisms in the next chapter.

REFERENCES

[1] Caenn R, Darley HCH, Gray GR. Composition and properties of drilling and completion fluids. 6th ed. Amsterdam: Elsevier; 2011.

[2] Fink J. Petroleum engineer's guide to oilfield chemicals and fluids. Amsterdam: Elsevier; 2012.

[3] Majidi R, Miska SZ, Ahmed R, Yu M, Thompson LG. Radial flow of yield-power-law fluids: numerical analysis, experimental study and the application for drilling fluid losses in fractured formations. J Pet Sci Eng 2010;70:334—43.

[4] Tehrani A. Behaviour of suspensions and emulsions in drilling fluids. Annu Trans Nord Rheol Soc 2007;15.

[5] Bourgoyne Jr AT, Millheim KK, Chenevert ME, Young Jr FS. Applied drilling engineering. Richardson: Society of Petroleum Engineers; 1991.

[6] Amani M, Al-Jubouri M, Shadravan A. Comparative study of using oil-based mud versus water-based mud in HPHT fields. Adv Pet Explor Dev 2012;4(5):18—27.

[7] Rehm B, Schubert J, Haghshenas A, Paknejad AS, Hughes J, editors. Managed pressure drilling. Houston, Texas: Gulf Publishing Company; 2008.

[8] Growcock FB, Patel AD. The revolution in non-aqueous drilling fluids. Paper AADE-11-NTCE-33 presented at the 2011 AADE National Technical Conference and Exhibition held at the Hilton Houston North Hotel, Houston, Texas, April 12—14, 2011.

[9] Shook CA, Roco MC. Slurry flow. Butterworth-Heinemann; 1991.

[10] Kim S, Karrila SJ. Microhydrodynamics: principles and selected applications. Butterworth-Heinemann; 1991.

[11] Kelessidis VC, Mpandelis G. Measurements and prediction of terminal velocity of solid spheres falling through stagnant pseudoplastic liquids. Powder Technol 2004;147:117—25.

[12] Hariharaputhiran M, Subramanian RS, Campbell GA, Chhabra RP. The settling of spheres in a viscoplastic fluid. J Non-Newtonian Fluid Mech 1998;79:87—97.

[13] Beris AN, Tsamopoulos JA, Armstrong RC, Brown RA. Creeping motion of a sphere through a Bingham plastic. J Fluid Mech 1985;158:219—44.

Chapter 5

Mechanisms and Diagnostics of Lost Circulation

Mud losses can be classified according to their severity as follows [1]:

- seepage losses (loss rate less than 10 bbl/h, ie, below 1.6 m^3/h);
- partial losses (10—100 bbl/h, ie,1.6—16 m^3/h);
- severe losses (more than 100 bbl/h, ie, above 16 m^3/h);
- total (no returns to surface).

The range of partial losses is sometimes defined from 10 to 500 bbl/h, and severe above 500 bbl/h [2]. Sometimes different classifications are used for water-base muds (WBMs) and oil-base muds (OBMs); for instance [3]:

- seepage losses
  - OBM: less than 10 bbl/h (1.6 m^3/h);
  - WBM: less than 25 bbl/h (4 m^3/h);
- moderate losses
  - OBM: 10 to 30 bbl/h (1.6—4.8 m^3/h);
  - WBM: 25 to 100 bbl/h (4—16 m^3/h);
- severe losses
  - OBM: more than 30 bbl/h (4.8 m^3/h);
  - WBM: more than 100 bbl/h (16 m^3/h);
- total (no returns to surface).

Seepage losses are common in high-porosity high-permeability rocks such as sandstone. Partial losses are common in unconsolidated sand or gravel. They can also be associated with narrow fractures (natural or induced). Severe losses are common in unconsolidated sand or gravel, too. They can also be associated with wider fractures (natural or induced). Total losses are often associated with vugular or cavernous formations, heavily fractured rocks, or fracture systems with large fracture apertures.

Seepage losses may be erroneously attributed to mud escape into the formation, while in reality they are caused by increasing hole volume [3]. As an example, consider a well 31 cm in diameter drilled with a 10 m/h penetration rate. The new hole volume created per hour is 0.75 m^3. This new volume needs to be filled with the drilling fluid. In addition, drilling fluid is retained on drill cuttings. As a rule of thumb, the amount of fluid retained on the cuttings is

Lost Circulation. http://dx.doi.org/10.1016/B978-0-12-803916-8.00005-4

equal to the volume of cuttings [3]. This is 0.75 m^3, in our example. Thus, the amount of drilling fluid "loss" caused by hole penetration is 1.5 m^3/h. This can easily be misinterpreted as seepage loss.

Classification of losses based on severity is convenient, but does not say much about the loss mechanism or about quantitative properties of, for example, fractures causing lost circulation. Preventing and curing mud losses calls for a good grasp of their mechanisms and the underlying physics. Different mechanisms may be at play in different formations. By their origin, losses are usually attributed to:

- porous matrix;
- large cavities (vugs);
- natural fractures;
- drilling-induced fractures.

These mechanisms are considered in detail in this chapter. After that, losses in different environments such as deepwater wells, depleted reservoirs, etc. are discussed. The chapter concludes with a discussion on loss detection and identification of loss mechanisms using different available downhole and surface measurements.

5.1 MEASURING THE LOSSES

When discussing the mechanisms of mud losses, we will refer to the *mud loss flow rate* as the flow rate of the fluid going into the well minus the flow rate of the fluid coming out of the well. This parameter can be measured, in practice, by installing flowmeters at the rig site. Mud losses caused by different mechanisms—eg, vugs and fractures—usually cause different flow rate signatures, ie, mud loss flow rate as a function of time. High-frequency measurements of mud loss flow rate provide a detailed account of losses. Electromagnetic flowmeters have been used to acquire high-frequency mud loss data, with the sampling rate of 0.2 s^{-1}. The accuracy of electromagnetic flowmeters used in Refs. [4,5] was 10−15 L/min.

The flowmeter measuring the flow into the well is installed on the standpipe. The flowmeter measuring the flow out of the well is installed upstream of the shakers. The operation of electromagnetic flowmeters requires that the drilling fluid is electrically conductive. This restricts the application of such devices to WBMs.

In practice, the electromagnetic flowmeters provide reliable data when measuring the flow rate into the well. The flow rate out of the well is measured less accurately since the mud coming out of the well has poorer quality and is contaminated with drill cuttings. Installation of electromagnetic flowmeters requires modifications to the mud flow system, which impairs their wide use. Instead of installing a flowmeter on the line in, pump stroke can be used to estimate the flow rate into the well [4,5].

Another parameter that can be monitored for mud loss diagnostics is the pit level. Monitoring the pit level by means of acoustic or floating sensors provides the cumulative volume of mud lost over a period of time. A drop in the mud-pit level can indicate a massive loss of circulation. The accuracy of these measurements, however, may be insufficient for detecting smaller incidents. Also, the time resolution does not allow pinpointing individual fractures. In addition, the pit level is affected by surface losses, filling and draining of surface lines, additions of water and chemicals to the mud, mud expansion/contraction cycles due to temperature and pressure variations in the hole, and other factors. Finally, on semisubmersible rigs, the pit level is affected by wave action, even though the time scales of mud losses are typically different from the period of wave motion [6]. This necessitates the use of high-frequency flow rate measurements in order to detect and interpret mud-loss incidents timely and rightly.

In order to improve detection of lost circulation and diagnostics of loss mechanisms, the flow rate and cumulative loss measurements should be supplemented by additional data:

- information about the lost circulation event (duration, depth, ECD);
- drilling fluid properties (mud weight, rheology, solids content, particle size);
- operational drilling parameters (weight on bit, torque, mud weight, standpipe pressure, rate of penetration, predicted fracture gradient);
- logging data.

We will return to mud loss measurements and diagnostics in Section 5.15 after studying the mechanisms of lost circulation.

5.2 LOSSES CAUSED BY HIGH-PERMEABILITY MATRIX

Losses into porous matrix occur when a high-permeability rock becomes exposed to the drilling fluid. Their onset is gradual, and the loss rate increases as more rock is exposed at the borehole wall as drilling proceeds. Gradually, filter cake builds up on the borehole wall and the loss rate starts to decrease. Losses continue at the drill bit until the high-permeability zone has been passed. This scenario produces the signature of mud loss rate versus time shown in Fig. 5.1A [5]. The variation of the pit level is shown in Fig. 5.1B.

Losses into porous matrix occur in unconsolidated formations (sands, gravel) and depleted reservoirs. The pore size needs to be larger than about 3 times the solid particle diameter in order for the mud to be able to enter the pore space [5].

Losses in highly fractured (defragmented) rocks can be considered as a variety of losses into high-permeability porous matrix. Such losses occur, for example, when drilling subsalt formations. Rubble zones are common at the bottom of the salt and may cause severe losses on exiting the salt.

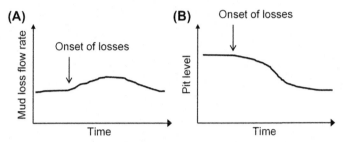

FIGURE 5.1   Mud losses into a high-permeability rock. Loss rate versus time (A) and pit level versus time (B).

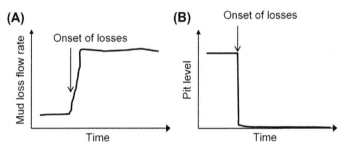

FIGURE 5.2   Mud losses in a cavernous/vugular formation: Schematic view of loss rate versus time (A) and pit level versus time (B).

5.3 LOSSES IN VUGULAR FORMATIONS

Vugular zones are found in rocks subject to dissolution over geologic time. Examples of such rocks are limestone, dolomite, and rock salt. Vugs can vary in size, sometimes taking the form of underground caves eagerly explored by speleologists and general public.

Losses start suddenly when the drill bit hits the vug. The cumulative volume of the lost fluid can be quite large, especially when several vugs are connected into one large system. Total losses may occur in such formations. A possible mud loss signature is shown in Fig. 5.2.

5.4 LOSSES CAUSED BY NATURAL FRACTURES

Natural fractures—ie, fractures that exist before the well is drilled—may provide pathways for the drilling fluid. According to one study, natural fractures were responsible for 76% of losses at one major operator [6].

Whether a natural fracture will cause mud losses depends on several factors:

- the wellbore pressure;
- the hydraulic aperture of the fracture;

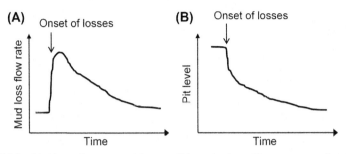

FIGURE 5.3 Mud losses into a natural fracture: Schematic view of loss rate versus time (A) and pit level versus time (B).

- the drilling fluid rheology and composition;
- fluid leakoff through the fracture walls and buildup of the filter cake inside the fracture.

To cause losses, the fracture needs to be wide enough and provide sufficient permeability for the drilling fluid. It follows that not every natural fracture will cause mud losses since fractures are, in many cases, closed or filled with gouge or mineralization products which make them too narrow or impermeable. If fractures are opened and make up a connected network, their capacity for accepting the drilling fluid can be virtually unlimited and may lead to severe or total losses [7]. Some lost circulation materials (LCMs) successfully used to treat losses caused by induced fractures are ineffective in naturally fractured rocks, because the permeability and volume of natural fracture networks may be much larger.

Mud losses into a natural fracture typically start with a sudden drop in the flow rate out of the well (Box 5.1). The mud loss flow rate versus time curve shows a sudden increase and gradual decrease in the differential flow rate, as illustrated in Fig. 5.3. The tailing stage of the mud loss event can be approximated by a square-root function of time, as pointed out by Sanfillippo et al. [4] in their analysis of field data:

$$V_{\text{cum}} = V_0 + C\sqrt{t} \qquad [5.1]$$

where V_{cum} is the cumulative volume of the drilling fluid lost into the fracture from the beginning of the mud loss incident up to time t; V_0 is the spurt loss; and C is an empirical coefficient.

Two major factors controlling the effect of natural fractures on drilling operations are the hydraulic aperture of the fracture, w_h, and the bottomhole pressure (BHP, P_w). Table 5.1 provides a classification of possible scenarios based on these factors.

Let us consider the nine cases of Table 5.1 one by one.

BOX 5.1 Severe Losses During Drilling in a Naturally Fractured Carbonate Formation

An incident of severe losses caused by a natural fracture or a fracture system was described by Sanfillippo et al. in their SPE paper 38177, "Characterization of conductive fractures while drilling."

Losses occurred at 3620 m depth and lasted for several hours. A fracture inclined at 70 degrees to the wellbore axis was identified in the image log. In total, 14−18 m³ of drilling fluid was lost into the formation. The flow rate was measured at the inlet and outlet of the well, and the difference between the two flow rates plotted versus time looked similar to Fig. 5.3A, indicating that losses were indeed caused by a hydraulically opened natural fracture. The mud loss flow rate increased rapidly at the beginning, reaching a maximum at 20 s after the onset of losses. The differential flow rate at the maximum was equal to 1270 L/min. The flow rate gradually declined thereafter. This decline was attributed by Sanfillippo et al. to fracture plugging by deposited solid particles (the average size of solid particles in the drilling fluid was 50 µm). As discussed in this chapter, the decline could be also due to the yield-stress rheology of the drilling fluid, in which case the losses would eventually stop even if there were no leakoff through the fracture walls, and thus filter cake deposition could not play a significant role.

No LCM was used during this mud loss incident. Sanfillippo et al. noted, however, that the losses might be stopped more efficiently if coarser solid particles were present in the drilling fluid. Such particles could plug the fracture nearer the well, reducing the total volume of losses and the duration of the incident.

TABLE 5.1 Losses Caused by Natural Fractures: Classification of Scenarios Based on Fracture's Hydraulic Aperture, w_h, and Bottomhole Pressure (BHP)

	BHP $< P_p$	$P_p <$ BHP $<$ FRP	BHP $>$ FRP
$w_h < w_{c1}$	(A) No losses	(D) No losses	(G) Losses may occur if BHP is high enough to open the fracture for flow
$w_{c1} < w_h < w_{c2}$	(B) No losses	(E) Minor losses; stop by themselves	(H) Losses will occur and might not stop by themselves
$w_h > w_{c2}$	(C) No losses	(F) Losses will not stop unless LCM or other treatment is used	(I) Losses will occur and will not stop by themselves

Cases (A), (B), and (C)

If the bottomhole pressure (BHP) remains below the fluid pressure inside the fracture, P_p, there will be no losses. Thus, the necessary condition for the mud to flow into a natural fracture is that the BHP be greater than the fluid pressure inside the fracture:

$$P_w > P_p \qquad [5.2]$$

Eq. [5.2] thus provides the necessary condition for mud losses. If this condition is not met, losses cannot occur. This is the principle behind underbalanced drilling (UBD), which is one of the most effective loss prevention techniques (chapter: Preventing Lost Circulation).

Case (D)

If the fracture aperture is below a certain critical value, w_{c1}, the solids contained in the drilling fluid will block the fracture near its mouth. There will be no mud losses unless the wellbore pressure is so high that it opens the fracture, thereby increasing the hydraulic aperture sufficiently to enable flow (case (G), considered later in this section).

Case (E)

If the fracture aperture is above w_{c1} but below the second threshold value, w_{c2}, mud losses will occur, but they will stop as the mud propagates inside the fracture away from the wellbore. At least two mechanisms have been proposed to explain why losses will eventually stop in this case.

According to one mechanism advocated in Ref. [4], losses eventually stop because the fracture becomes plugged by solid particles present in the mud. This explanation is consistent with the experimental results of Ref. [6] suggesting that a WBM with leakoff through fracture walls produces consistently smaller mud loss volumes than a similar WBM without leakoff.

According to another view, advocated in Refs. [8−11], the stop in the mud propagation is attributed to the yield-stress rheology of the drilling fluid. A yield-stress fluid can only flow inside the fracture if the magnitude of the pressure gradient overcomes a critical value given by

$$|\nabla P|_c = 2\tau_Y/w \qquad [5.3]$$

where w is the fracture aperture; τ_Y is the yield stress of the fluid. When the yield-stress fluid flows, a flow region is created near the fracture walls, where the shear rate is maximum. Away from the walls, near the center plane, a plug region is found. The width of the plug region is given by

$$w_{\text{pl}} = 2\tau_Y/|\nabla P| \qquad [5.4]$$

where $|\nabla P|$ is the magnitude of the pressure gradient.

Consider the example of a vertical borehole that hits a horizontal natural fracture during drilling. If the BHP exceeds the fluid pressure inside the fracture, mud losses begin. The drilling fluid displaces the formation fluid toward the periphery of the fracture. The flow pattern is approximately radial, and the interface between the formation fluid and the drilling fluid gradually expands. If the BHP remains constant during this process, the fluid gradient inside the fracture will eventually drop everywhere below the critical value given by Eq. [5.3]. As it happens, the fluid will stop flowing. The reason for mud losses to stop, according to this mechanism, is the nonzero yield stress of the drilling fluid. If the drilling fluid were Newtonian—ie, $\tau_Y = 0$ the flow would persist as long as there is a pressure difference between the borehole and the formation.

According to a mud loss model proposed by Liétard et al. [8], the terminal radial penetration of a Bingham mud into a smooth-walled fracture of constant aperture, w ($w = w_h$ for such a fracture), is given by

$$r_{max} = r_w + w\Delta P/3\tau_Y \qquad [5.5]$$

where r_w is the radius of the well; $\Delta P = P_w - P_p$ is the overbalance (the wellbore pressure, P_w, minus the formation fluid pressure, P_p). The maximum propagation distance depends only on the yield stress; the plastic viscosity does not enter the equation. The plastic viscosity does affect, however, the *dynamics* of the mud loss event: the mud loss flow rate is higher if the mud has lower plastic viscosity.

In the case of a fluid described by the Herschel—Bulkley rheological model,

$$r_{max} = r_w + \frac{n+1}{2n+1}\frac{w\Delta P}{3\tau_Y} \qquad [5.6]$$

where n and τ_Y are the flow behavior index and the yield stress of the Herschel—Bulkley fluid. Eq. [5.5] is a specific case of Eq. [5.6] when $n = 1$. The terminal radial position of the mud front is not affected by the consistency index of the Herschel—Bulkley fluid. The consistency index does affect, however, the *dynamics* of the mud loss event, namely the magnitude of the mud loss flow rate.

The total volume of the fluid lost into the fracture is given by:

$$V_{lost} = \pi w \left(r_{max}^2 - r_w^2 \right) \qquad [5.7]$$

Eq. [5.7] shows that the volume of mud lost into a natural fracture is indeed finite and governed by the fracture aperture, the overpressure, and the rheological properties of the drilling fluid. Thus, the yield-stress rheology can indeed explain finite losses into natural fractures. The other mechanism mentioned above is fracture bridging and sealing by deposited solid particles transported by the mud (clay, barite, drill cuttings). In reality, both mechanisms, ie, the fluid rheology and the particle deposition, are likely to be at play, in particular in high-permeability rocks where the filter cake deposition inside the fracture is facilitated by leakoff through the fracture walls.

The yield stress and the flow behavior index to be used in Eqs. [5.5] and [5.6] should be measured at the downhole conditions, ie, elevated temperature and pressure.

Eqs. [5.5]–[5.7] suggest that it might be possible to evaluate the fracture aperture during drilling if the total volume of lost fluid is measured during a mud loss incident. Additional insight into the dynamics of a mud loss incident can be gained by analyzing the mud loss flow rate as a function of time. For a Bingham fluid spreading radially inside a horizontal fracture, the radial distance of the mud front from the well axis, r, at time t can be evaluated. A closed-form solution for r exists, given implicitly by Refs. [8,12]:

$$
t_D = 4r_{Dmax}(r_{Dmax} - 1)\left\{ -\ln r_D\left[\frac{r_D}{r_{Dmax}} + \ln\left(1 - \frac{r_D}{r_{Dmax}} \right) \right] \\ + \sum_{n=2}^{\infty} \frac{1}{n^2}\left[\left(\frac{1}{r_{Dmax}} \right)^n - \left(\frac{r_D}{r_{Dmax}} \right)^n \right] \right\}
$$

[5.8]

where the dimensionless radial position of the mud invasion front, r_D, is defined as

$$
r_D = r/r_w
$$

[5.9]

with r_w being the radius of the well. The dimensionless time that has passed since the moment when the borehole hit the fracture, t_D, is defined as

$$
t_D = \left(\frac{w}{r_w} \right)^2 \frac{t\Delta P}{3\mu_{pl}}
$$

[5.10]

with μ_{pl} being the plastic viscosity of the Bingham fluid.

The dimensionless maximum mud invasion radius is given by:

$$
r_{Dmax} = r_{max}/r_w = 1 + \frac{1}{\alpha}
$$

[5.11]

where α is the so-called dimensionless finite invasion factor defined as (cf. Eq. [5.5]):

$$
\alpha = \frac{3r_w}{w} \frac{\tau_Y}{\Delta P}
$$

[5.12]

Application of Eq. [5.8] can be illustrated as follows. Consider a borehole of 30 cm in diameter, drilled with an overbalance of $\Delta P = P_w - P_p = 0.1$ MPa. The yield stress and the plastic viscosity of the drilling mud are equal to 5 Pa and 10 cP, respectively. Applying Eq. [5.8] for two values of the fracture aperture, $w = 0.5$ mm and $w = 1$ mm, yields the results presented in Figs. 5.4 and 5.5. The solid lines in Figs. 5.4 and 5.5 represent the case of $w = 0.5$ mm, ie, $\alpha = 0.045$. The dotted lines in Figs. 5.4 and 5.5 represent the case of $w = 1$ mm, ie, $\alpha = 0.0225$. The dimensionless volume of the lost fluid plotted in Fig. 5.4 is defined as

$$V_{Dlost} = \frac{V_{Dlost}}{\pi r_w^2 w} = r_D^2 - 1 \qquad [5.13]$$

Doubling the fracture aperture, from 0.5 to 1 mm, results in the mud front propagating approximately twice as far into the fracture in Fig. 5.5, in accord

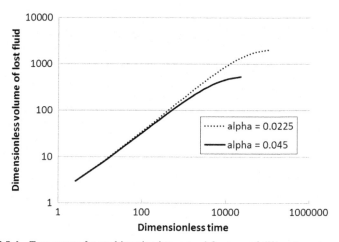

FIGURE 5.4   Type curves for mud invasion into natural fractures of different aperture.

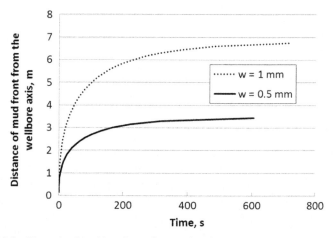

FIGURE 5.5   Dimensional mud invasion radius as a function of dimensional time in fractures of different aperture.

with Eq. [5.5] (the radius of the well is considerably smaller than the mud invasion radius in this example).

Fig. 5.4 suggests that not only the total amount of the fluid lost into the fracture (determined by $r_{D\text{max}}$) but also the shape of the curve "volume versus time" depends on the fracture aperture. Liétard et al. [8] obtained this result and suggested that the nondimensional "type curves" like the ones shown in Fig. 5.4 could be used to estimate the fracture aperture from mud loss data more accurately. In this case, the entire curve would be matched and not only the value of r_{max}. This approach was later elaborated for a Herschel−Bulkley fluid by Majidi et al. [9]. No closed-form solution of the flow problem is, however, available for Herschel−Bulkley fluid, which necessitates the use of numerical methods when constructing the type curves for this fluid.

The theoretical argument from which Eqs. [5.5]−[5.13] follow should be treated with caution since it makes use of several simplifying assumptions. The following assumptions were made in the models in Refs. [8,10]:

1. The fracture is a single isolated fracture, not a part of a fracture network.
2. The fracture aperture does not change as the pressure inside the fracture increases.
3. The formation fluid in place offers no resistance to the mud invading the fracture.
4. The formation fluid and the drilling fluid are incompressible.
5. There is no leakoff through the fracture wall and, consequently, no filter cake builds up in the fracture.
6. The flow is laminar.

The first assumption is not always realistic: as we have seen in chapter "Natural Fractures in Rocks," fractures are often part of complex three-dimensional connected networks. However, the model might still provide satisfactory results as long as the mud front did not reach the intersection between the invaded fracture and another fracture, or a branching point of the invaded fracture. What happens after that is pure guesswork. As pointed out by Liétard et al. [8], a connected fracture network creates more opportunities for the drilling fluid to escape. On the other hand, the flow divides at intersection points, and the flow rate thus decreases in individual fractures at each such division. This effectively increases the apparent fluid viscosity, making it more difficult for the fluid to propagate further along the fracture network. The resulting effect of fracture connectivity on mud losses depends on which of these two effects dominates.

The second assumption in the above list—ie, constant fracture aperture unaffected by the pressure increase in the fracture—holds true as long as the fluid pressure inside the fracture stays below the fracture reopening pressure (FRP). Under these conditions, the fracture remains mechanically closed (its faces are in touch with each other) even as the fluid pressure inside the fracture increases above the formation fluid pressure. If the fluid pressure overcomes the reopening pressure, fracture's hydraulic aperture will increase, which will reduce the resistance to flow and further promote the fluid loss and fracture reopening. This happens in cases (G), (H), and (I) in Table 5.1. Whether losses will eventually stop in this case depends on the amount of overpressure and the fracture stiffness. A theoretical model intended to capture this aspect of mud losses was developed in Refs. [13,14]. Laboratory experiments [10] revealed the strong sensitivity of mud losses to the fracture opening caused by the pressure increase inside the fracture.

The third and fourth assumptions in the above list imply that the fracture initially is empty or is filled with an ideal fluid. This might not introduce an appreciable error if the fracture is initially filled with gas, but will affect the mud flow into the fracture if the fracture is initially filled with a viscous fluid like water or oil. The assumption of an initially empty fracture was relaxed in Ref. [9]. If the mud has Bingham rheology, and the formation fluid is Newtonian, the effect of formation fluid resistance on the mud flow rate in the fracture becomes appreciable when the viscosity of the formation fluid is equal to (or higher then) the plastic viscosity of the mud. The final invasion radius is not affected by formation fluid properties and is solely a function of the overbalance, yield stress of the mud, and the fracture hydraulic aperture [9].

The fifth assumption—ie, the absence of leakoff through the fracture walls—might hold, at least approximately, at the time scale of a mud loss incident, and in some formations, eg, shale. In general, however, the "no-leakoff" assumption is not valid. Leakoff through fracture walls will result in a filter cake deposited on the fracture walls. This will reduce the mud invasion in the fracture. As pointed out in Ref. [6], leakoff increases the viscosity of the slurry flowing in the fracture, which tends to reduce (and possibly stop) the losses. On the other hand, the mud filtrate can flow past the filter cake and deeper into the fracture. This filtrate will have Newtonian rheology, or at least will be thinner than the original mud. It might therefore flow farther into the fracture than a yield-stress mud would if there were no leakoff through the fracture wall [8].

Even though practical applicability of lost circulation models developed in Refs. [8–10] is somewhat limited by the simplifying assumptions discussed above, the models provide valuable insights into the mechanics of lost circulation.

Unlike induced fractures, natural fractures may have any orientation with respect to the borehole. In particular, a natural fracture may be oriented in such way that would facilitate slip (shear movement) along the fracture plane when the drilling fluid invades the fracture. When mud invasion happens, the effective normal stress acting on the fracture plane drops, and so does the frictional resistance which is proportional to the effective normal stress. This may initiate shear displacement if the in situ stresses exert sufficient shear stress in the fracture plane [15]. Such shear displacement may eventually lead to stuck pipe, which will additionally aggravate the lost-circulation problem.

An order-of-magnitude estimate of the shear displacement, ΔL, is given by [16]

$$\Delta L = \frac{1 - v^2}{E} R_f \tau \qquad [5.14]$$

where E and v are the Young's modulus and the Poisson's ratio of the rock; R_f is the radius of the fracture; τ is the in situ shear stress acting in the fracture plane. From Eq. [5.14], the magnitude of the shear displacement increases with the dimensions of the fracture and with the shear stress. Thus, larger contrast between in situ principal stresses may produce larger (thus more dangerous) shear displacements caused by mud losses. Regarding the fracture dimensions, it is estimated that borehole wall shifts become significant when the fracture radius is larger than 15 m [16].

Case (F)

One problem with the rheological argument presented under Case (E) is that, according to Eq. [5.7], mud losses into natural fractures will always stop because the value of V_{lost} is always finite. Empirical evidence suggests that mud losses do not stop by themselves if the fracture aperture exceeds a certain critical value, w_{c2}. If the fracture aperture is above w_{c2}, special measures, such as injection of an LCM, are required to stop the losses. The values of w_{c1} and w_{c2} depend on the composition and rheology of the drilling fluid as well as on the rock properties. For example, empirical data provided in Ref. [6] suggest that losses do not occur as long as the fracture aperture is below $w_{c1} = 150 \div 250\,\mu m$. At such low apertures, barite particles $(0-100\,\mu m)$ cannot enter the fracture [17,18]. Detectable but limited mud losses occur when the fracture is between $w_{c1} = 150 \div 250\,\mu m$ and $w_{c2} = 500 \div 750\,\mu m$ wide. Mud losses occur and do not stop unless LCM is used if the fracture aperture is greater than $w_{c2} = 500 \div 750\,\mu m$.

Fractures of different apertures thus have unequal potential to cause losses. A 1-mm-wide fracture may cause more trouble than a fractured zone containing 10 fractures, each 0.1 mm wide. When it comes to lost circulation, properties of individual fractures, in particular their apertures, are more important than the number or density (ie, spacing) of the fractures.

So far in this Section, we have been concerned with losses into natural fractures occurring without fracture opening, ie, while the fracture aperture remains constant. If the borehole pressure becomes sufficiently high, a preexisting, natural fracture may start reopening (the term "*re*opening" here indicates that the *first* opening occurred when the fracture was created long time ago in geological past). Reopening of a natural fracture exposed to the drilling fluid can be thought of as a process somewhat similar to the repeat cycle of an extended leakoff test (XLOT). The mechanics of fracture reopening depends on the degree of hydraulic communication between the fracture and the wellbore. With regard to the hydraulic communication, the following scenarios are possible:

1. (H) and (I) in Table 5.1: The hydraulic fracture aperture is so large that the drilling fluid would be able to enter the fracture even without reopening, as soon as the wellbore pressure becomes greater than the fluid pressure inside the fracture. "So large" here means that the initial fracture aperture is greater than the threshold value, w_{c1}. The fracture aperture can be large; eg, because of asperities holding the fracture faces apart or because the effective normal stress acting on the fracture plane is too small. For a sufficiently wide fracture aperture, the fluid pressure at the mouth of the fracture will then be equal to the wellbore pressure at all times.

2. (G) in Table 5.1: The fracture has such small aperture that the drilling fluid would not be able to enter the fracture without reopening. The small aperture prevents fluid pressure communication between the wellbore and the fracture, with significant consequences for the fracture reopening process.

The *fracture mouth* mentioned above is the part of the fracture in the immediate vicinity of the well. The *fracture tip* (or *fracture front*), in contrast, is the part of the fracture farthest away from the well, where the fracture will advance if it starts propagating.

Let us consider cases (H) and (I) first.

Cases (H) and (I)

In these cases, there is perfect hydraulic communication between the well and the fracture. These cases are similar to the case usually considered when treating the reopening process in hydraulic fracturing. The theory developed for hydraulic fracturing can easily be adapted to lost circulation in Cases (H) and (I). Consider the instant when the borehole hits the fracture during drilling. The fluid pressure at the intersection between the borehole and the fracture (the "fracture mouth") rapidly increases from the formation fluid pressure, P_p, to the wellbore pressure, P_w. If the magnitude of the wellbore pressure is greater than the rock stress acting normal to the fracture plane at the fracture mouth, the fracture will start opening at the mouth—ie, its local aperture, w—will increase there. The drilling fluid will flow into the fracture, displacing the formation fluid, which is similar to Cases E and F discussed above. The difference here is that the fracture aperture will increase as the fluid enters the fracture. The normal stress that needs to be overcome in order to open the natural fracture at the mouth depends on the stress concentration in the near-well area.

As the drilling fluid flows into the fracture and the fluid displacement front exits the near-well stress concentration zone, the normal stress acting on the fracture plane at the mud front becomes equal to the in situ stress in the direction normal to the fracture. The fluid pressure inside the fracture far away from the well must thus be sufficiently high to overcome this in situ stress (plus frictional pressure losses along the fracture) in order to reopen the fracture further.

The fracture reopening is a *gradual process*, as illustrated in Figs. 5.6 and 5.7. As such, the very concept of the FRP as a single value of the wellbore pressure at which the fracture "reopens," is therefore somewhat misleading. The fracture reopening takes place within a range of wellbore pressure values. Reopening starts at the fracture mouth and propagates along the fracture.

As the reopening proceeds, the fracture aperture increases, which facilitates the fluid flow into the fracture by increasing the fracture permeability.

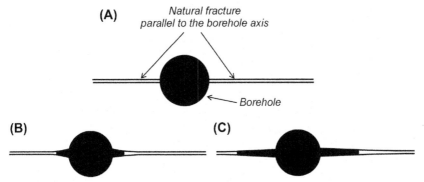

FIGURE 5.6   Reopening of a natural fracture in Cases (H) and (I): (A) before fluid starts entering the fracture; (B) drilling fluid is entering the fracture, and the fracture reopens near the mouth; (C) the drilling fluid continues entering and reopening the fracture farther away from the mouth. The fracture plane is parallel to the borehole axis in this example. The fracture faces are idealized (planar and smooth). Drilling fluid is shown as *black fill*.

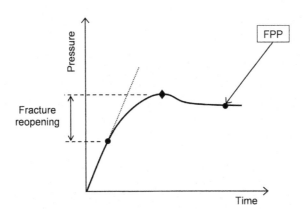

FIGURE 5.7   Schematic illustration of the ambiguity in defining the fracture reopening pressure. The plot shows the wellbore pressure versus time curve when a preexisting fracture is pressurized. *FPP*, fracture propagation pressure.

At the same time, the same mechanism that limits mud losses into natural fractures without reopening, namely the effect of the fluid yield stress, is at work here. The relative effect of these two mechanisms will determine whether the losses will eventually stop, and how much fluid will be lost by that time. The cumulative losses will obviously be larger in Cases (H) and (I) than if the wellbore pressure were insufficient to reopen the fracture. If the wellbore pressure approaches the in situ stress acting normal to the fracture plane, the fracture opening will increase significantly. If there were no pressure drop inside the fracture caused by friction at the fracture walls, the fracture could become very wide as the wellbore pressure equals the in situ stress normal to fracture. This might entail total loss of circulation, with all drilling mud disappearing into the fracture. From this viewpoint, the most unfavorable orientation of the natural fracture is when the fracture plane is oriented normal to the minimum in situ stress. This is one of the reasons why the minimum in situ stress is often adopted as the upper bound of the operational pressure window. Natural fractures having orientations other than normal to the minimum in situ stress will have higher FRP and thus will be able to sustain higher BHPs without severe losses. This fact is well recognized in the drilling and rock mechanics communities [19]. In reality, due to frictional pressure losses inside the fracture and (in permeable rocks) due to leakoff through the fracture wall, a wellbore pressure *higher* than the in situ stress normal to fracture would be required to reopen the entire natural fracture.

Now let us consider some more specific examples of fracture reopening with full hydraulic communication between the borehole and the fracture. The near-well stress concentration that must be overcome in order to reopen the fracture at the mouth, depends on the mutual orientations of the in situ stress tensor, the borehole, and the fracture plane. Consider a vertical borehole drilled in a formation with extensional stress regime, ie, $\sigma_v > \sigma_H > \sigma_h$. A natural fracture met by the borehole during drilling can have any orientation. As our first example, consider a planar fracture oriented perpendicular to the minimum in situ stress, σ_h (and, thus, parallel to the borehole axis). Assume that the wellbore is drilled such that its axis lies in the fracture plane. The fracture intersects the borehole wall along two straight lines. Neglecting poroelastic and thermal effects in the near-well area and assuming elastic deformation of rocks around the well, the total hoop stress keeping the fracture closed at the mouth is given by $(3\sigma_h - \sigma_H - P_w)$ where P_w is the fluid pressure in the wellbore (the BHP). If the horizontal stress anisotropy is so large that $\sigma_H > 3\sigma_h$, the fracture will be opened at the mouth no matter how low the wellbore pressure is. Assume therefore that $\sigma_H < 3\sigma_h$. In this case, which is realistic in most fields, the fracture will start opening at its mouth when the fluid pressure inside the fracture, near the mouth, exceeds $(3\sigma_h - \sigma_H - P_w)$. Since in Cases (H) and (I), the fluid pressure in the fracture near its mouth is equal to the wellbore pressure, the fracture will start opening when the well-bore pressure exceeds the following value [20]:

$$\text{FRP}_{\text{start}} = \frac{1}{2}(3\sigma_h - \sigma_H) \qquad [5.15]$$

Since $\sigma_H > \sigma_h$, the value of $\text{FRP}_{\text{start}}$ given by Eq. [5.15] is lower than the minimum in situ stress. Recall now that the natural fracture becomes opened along its entire length when the fluid pressure inside the fracture becomes equal to the normal in situ stress acting normal to the fracture plane. Hence, reopening of the fracture considered in this example occurs within a range of wellbore pressures from $P_w = (3\sigma_h - \sigma_H)/2$ to $P_w = \sigma_h$. The upper bound of this range will be, indeed, larger than σ_h because of fluid pressure losses due to friction inside the fracture. The range of reopening pressure values becomes more narrow if the stress anisotropy decreases; ie, when σ_H becomes closer to σ_h. When $\sigma_H = \sigma_h$, the FRP is equal to $\sigma_h = \sigma_H$, according to Eq. [5.15].

As mentioned above, the pore pressure change in the near-well area and poroelastic effects associated with this change were neglected when deriving Eq. [5.15]. Neglecting the pore pressure redistribution in the near-well zone is a reasonable assumption when considering lost circulation. In a low-permeability rock, it would take some time for the pore pressure to in-crease. In a high-permeability rock, the filter cake deposited on the borehole wall might serve to protect the pressure diffusion from the well into the for-mation. Since lost circulation normally starts during drilling, the delayed buildup of the poroelastic contribution to the hoop stress will not be so important. In situations where this effect cannot be neglected, the poroelastic contribution needs to be added to the hoop stress. Assume that there is no filter cake on the borehole wall, and the rock has such high permeability that the pore pressure is equal to the wellbore pressure near the wall. Under these assumptions, the upper bound of the poroelastic contribution to the hoop stress concentration is given by $\alpha(1 - 2v_d)(P_w - P_{p0})/(1 - v_d)$ where P_{p0} is the initial pore pressure (before drilling). This yields the following estimate of the wellbore pressure at which the fracture reopening starts at the mouth:

$$\text{FRP}_{\text{start}} = \frac{3\sigma_h - \sigma_H - \alpha\frac{1-2v_d}{1-v_d}P_{p0}}{2 - \alpha\frac{1-2v_d}{1-v_d}} \qquad [5.16]$$

If $\alpha = 0$ (no poroelastic effects), Eq. [5.16] reduces to Eq. [5.15]. The value of $\text{FRP}_{\text{start}}$ given by Eq. [5.16] is greater than the one given by Eq. [5.15] if the initial pore pressure is lower than $(3\sigma_h - \sigma_H)/2$.

As another example, consider a horizontal fracture in the same stress environment, ie, $\sigma_v > \sigma_H > \sigma_h$. Assuming that the borehole is vertical and very long, the in situ stress acting to close the fracture is on the order of σ_v along the entire fracture length. The FRP is equal to σ_v in this case, which can be considerably larger than the maximum reopening pressure for a vertical fracture considered above.

Finally, consider a natural vertical fracture oriented perpendicular to σ_H. Retracing the argument developed above for a fracture normal to σ_h, such fracture will start opening at its mouth when the wellbore pressure exceeds

$$\text{FRP}_{\text{start}} = \frac{1}{2}(3\sigma_H - \sigma_h) \qquad [5.17]$$

The wellbore pressure required to open the *entire* fracture in a static fashion (no flow, thus no friction-induced pressure losses) is equal to σ_H in this case. In reality, the upper bound of the reopening pressure will be higher than σ_H because a mud loss event is a dynamic process, and a pressure loss will occur inside the fracture. The lower bound of the reopening pressure interval may be lower than $\text{FRP}_{\text{start}}$ given by Eq. [5.17] because Eq. [5.17] is based on the assumption of linear elastic rock deformation around the well.

If the fracture plane is oblique—ie, neither normal nor parallel to the borehole axis—the fracture mouth is represented by an elliptical trace of the fracture on the borehole wall. Reopening of such a fracture will start not at a single $\text{FRP}_{\text{start}}$ value but within a range of $\text{FRP}_{\text{start}}$ values since the normal stress is different at different azimuthal locations along the fracture intersection with the borehole wall.

Since the orientation of natural fractures is not known beforehand, it is impossible to predict reliably whether losses will occur while drilling through a naturally fractured rock. The estimates of $\text{FRP}_{\text{start}}$ given above only indicate the wellbore pressure at which the fracture *starts* opening at the mouth. Opening at the mouth does not mean that there will be loss of drilling fluid in excess of what would be lost solely because of the pressure difference between the well and the fracture.

Opening of the entire fracture when the wellbore pressure approaches the upper end of the reopening pressure interval will cause more trouble since the entire fracture can accept significant amount of mud at that pressure. The value of the pressure at which this happens depends on the in situ stresses and the orientation of the natural fracture. The worst-case scenario is a natural fracture normal to σ_h, in which case the upper bound of the reopening pressure interval is equal to σ_h (plus frictional pressure drop inside the fracture). This value is often chosen as the upper operational pressure bound.

Case (G)

Now let us turn our attention to the last case, Case (G). Since the fracture aperture at the mouth is smaller than w_{c1} in this case, the fracture initially is effectively sealed at the mouth, and the fluid pressure in the entire fracture is approximately equal to the formation fluid pressure. The latter can be affected by poroelastic effects and pore pressure diffusion in the near-well area. For the sake of simplicity, we neglect these effects here, and assume that the formation fluid pressure is constant and equal to the initial pore pressure in the formation, $P_p = P_{p0}$.

Thus, in order for the fluid to start entering the fracture, the fracture aperture must be increased at least above w_{c1} at the mouth. The magnitude of the wellbore pressure at which this happens depends on the fracture orientation.

As a specific example, consider a planar fracture oriented perpendicular to the minimum in situ stress, σ_h (and, thus, parallel to the borehole axis). Assume that the wellbore is drilled so that its axis lies in the fracture plane, and the fracture therefore intersects the borehole wall along two straight lines. Neglecting poroelastic and thermal effects in the near-well area and assuming elastic deformation of rock around the well, the total hoop stress keeping the fracture closed at the mouth is given by $(3\sigma_h - \sigma_H - P_w)$ where P_w is the fluid pressure in the well. Assume that $\sigma_H < 3\sigma_h$ (otherwise the fracture would be opened at the mouth). The fracture will start opening at the mouth when the fluid pressure inside the fracture, near the mouth, exceeds $(3\sigma_h - \sigma_H - P_w)$. Since in Case (G) the fluid pressure inside the fracture is initially equal to the formation fluid pressure, the fracture will start opening at the mouth when the wellbore pressure exceeds the following value [20]:

$$\text{FRP}_{\text{start}} = 3\sigma_h - \sigma_H - P_p \qquad [5.18]$$

The value of $\text{FRP}_{\text{start}}$ given by Eq. [5.18] is greater than the value of $\text{FRP}_{\text{start}}$ given by Eq. [5.15] as long as $P_p < (3\sigma_h - \sigma_H)/2$. This demonstrates that the sealing capacity of the drilling fluid is essential for reducing mud losses in naturally fractured rocks.

Raising the wellbore pressure above the value given by Eq. [5.18] will not necessarily bring about mud losses. Firstly, the drilling fluid will not necessarily enter the fracture since the fracture aperture might still remain below w_{c1} at the mouth after the wellbore pressure has exceeded $\text{FRP}_{\text{start}}$. Secondly, even if the drilling fluid enters the fracture, the fluid penetration into the fracture may be quite limited since the fluid can only enter those parts of the fracture where the fracture aperture becomes greater than w_{c1}. These are the reasons why Eq. [5.18] only provides the most pessimistic value of the FRP for a drilling fluid with sealing capacity, as pointed out by Morita et al. [21].

In order to push the drilling fluid deeper into the fracture—ie, farther away from the well—the wellbore pressure must be increased further, which will reopen additional parts of the fracture. This process is illustrated in Fig. 5.8.

The value of the wellbore pressure that is required to reopen the natural fracture to a given extent and inject the drilling fluid into the fracture up to some distance from the well depends on the following factors [21,22]:

- orientation of the natural fracture;
- in situ stresses;

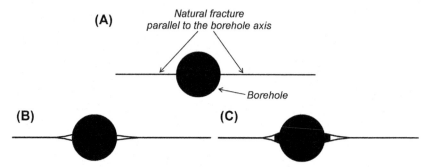

FIGURE 5.8 Reopening of a natural fracture in Case (G) (thin fracture sealed at the mouth): (A) before reopening starts; (B) the fracture has reopened near the mouth, but the drilling fluid has not entered the fracture; (C) the fracture has opened further, and the drilling fluid has entered part of the fracture. The fracture plane is parallel to the borehole axis in this example. The fracture faces are idealized (planar and smooth). Drilling fluid is shown as *black fill*.

- cooling of the near-well area by the drilling fluid;
- borehole orientation;
- borehole diameter;
- Young's modulus of the rock;
- permeability of the rock;
- capacity of the drilling fluid to build a filter cake and seal the fracture.

The theory developed by Morita et al. [21] in their pioneering work on the theory of lost circulation enables one to estimate the lost-circulation pressure if detailed knowledge of the above factors is available. Unfortunately, this is rarely the case in real life. A second shortcoming of the analysis presented in Refs. [21,22] is that it is quasi static; ie, the actual flow and displacement of the formation fluid by the drilling fluid are not considered.

Notwithstanding some limitations of the analysis by Morita et al. [21], it provides valuable insights into the effect of different operational and in situ parameters on mud losses in case of natural fractures of small aperture.

The effect of the fracture orientation can be investigated by considering a fracture that is oriented normal to the intermediate in situ stress, σ_H. The wellbore is again assumed to be vertical and parallel to the maximum in situ stress, σ_v. The fracture will start reopening at the mouth when the wellbore pressure exceeds

$$\text{FRP}_{\text{start}} = 3\sigma_H - \sigma_h - P_p \qquad [5.19]$$

which is greater than the value of $\text{FRP}_{\text{start}}$ for a fracture normal to σ_h, given by Eq. [5.18]. The fracture normal to σ_h represents the worst-case orientation of a natural fracture, from the lost circulation perspective. Eqs. [5.18] and [5.19] clearly demonstrate the effect of in situ stress magnitudes on the fracture reopening process (see also Box 5.2).

BOX 5.2 Effect of Hydraulic Communication Between Wellbore and Natural Fracture on the Fracture Reopening

A natural fracture starts reopening at the mouth when the wellbore pressure becomes greater than a certain threshold value, FRP_{start}. The value of FRP_{start} is affected by many factors; eg, in situ stresses, cooling of the near-well area, etc. One of the key factors affecting FRP_{start} is the degree of hydraulic communication between the fracture and the wellbore. Consider an example where a vertical fracture is drilled in a formation with the following in situ stresses:

$\sigma_v = 30$ MPa

$\sigma_H = 24$ MPa

$\sigma_h = 22$ MPa

With regard to the formation pore pressure, let us consider two cases:

Case (i): $P_p = 12$ MPa

Case (ii): $P_p = 20$ MPa

Consider a natural fracture oriented perpendicular to the minimum in situ stress, σ_h, and intersecting the well.

If full hydraulic communication is established between the well and the fracture, the fracture will start reopening at the mouth when the wellbore pressure becomes greater than $FRP_{start} = $ **21 MPa**, in both Cases (i) and (ii) (Eq. [5.15]).

If there is no hydraulic communication between the well and the fracture—ie, the fracture is sealed at the mouth—the value of FRP_{start} is equal to **30 MPa** in Case (i) and equal to **22 MPa** in Case (ii) (Eq. [5.18]).

This highlights the effect of fracture sealing on the reopening pressure.

It should be remembered, however, that fracture reopening at the mouth does not imply an immediate fluid loss into the fracture. Thus, a *lower* value of FRP_{start} does not necessarily mean that the borehole is more prone to lost circulation or that losses will be more severe in such well.

Cooling of the borehole wall by circulating drilling fluid reduces the hoop stress and thereby facilitates fracture reopening.

Reducing the borehole diameter increases the FRP, thereby suppressing mud losses. Reducing the Young's modulus of the rock facilitates the fracture reopening and thereby facilitates mud losses.

The capacity of the drilling fluid to build filter cake and seal the natural fracture plays an important role in fracture reopening. A drilling fluid that can seal a wider fracture will have a higher FRP. Indeed, in this case the fracture can reopen wider (while still being sealed) before the seal is broken and the fluid enters deeper into the fracture. For this reason, WBMs produce smaller mud losses than OBMs do.

The theoretical concepts presented in the nine Cases (A) through (I) can be summarized as follows:

If the aperture of a natural fracture is smaller than a certain threshold level, w_{c1}, the wellbore pressure in excess of the pore pressure is not sufficient to induce mud losses. The wellbore pressure must exceed a certain value above which the fracture reopens at the mouth, and it becomes possible for the drilling fluid to enter the fracture. Even in this case the losses still may be quite limited if the drilling fluid exhibits good sealing properties. Even if the latter is not the case, the losses still may be quite limited due to yield-stress rheology of the drilling fluid, which makes the fluid propagation into the fracture stop after a while. This shows that a natural fracture with the aperture smaller than w_{c1} should not cause much trouble in terms of lost circulation. Moreover, estimating the lost-circulation pressure based on the hoop stress concentration or on the value of the minimum in situ stress will produce a rather conservative upper pressure bound in this case. The well can in reality sustain higher pressures than the upper pressure bound obtained from the hoop stress analysis.

If the aperture of the fracture is above w_{c1} but below w_{c2}, the drilling fluid can enter the fracture if the wellbore pressure exceeds the formation fluid pressure inside the fracture. The drilling fluid can then flow inside the fracture, displacing the formation fluid, even if the fracture is mechanically closed; ie, its faces are in contact with each other at some spots (due to asperities). The fracture is thus *hydraulically* opened. If the wellbore pressure is not sufficiently high to open the fracture *mechanically*, the losses will eventually stop after the fluid front has propagated far enough from the well. This termination of losses is due to yield-stress rheology of the drilling fluid. The losses will be limited in this case. If the wellbore pressure is sufficiently high to reopen the fracture—ie, to increase its aperture—the amount of losses will be larger. Whether the losses will eventually stop in this case, will depend on the amount of overbalance, the Young's modulus of the rock, and the rheological properties of the drilling fluid.

If the aperture of the natural fracture is above w_{c2} (for example, a fracture located at shallow depth and having very rough faces), the fluid losses cannot stop by themselves, even if the wellbore pressure is not sufficient to cause fracture reopening. The fluid rheology alone cannot bring the losses under control in this case, and LCM is needed to stop the losses.

The values of w_{c1} and w_{c2} depend on the type of the drilling fluid, in particular its rheological properties and solids content. For a Newtonian drilling fluid (water or brine), w_{c1} is equal to zero. WBM usually has greater w_{c1} than OBM does because WBM has better filtration properties and more effective filter cake buildup in the fracture, which results in WBM's ability to seal wider fractures. Therefore, manipulating rheological and sealing properties of drilling fluids is one of the most effective ways to reduce mud losses. The theory presented in this section creates the foundation for some loss prevention methods discussed in chapter "Preventing Lost Circulation."

Note that natural fractures considered in this section were assumed to have infinite (or very large, in practice) in-plane dimensions. Such fractures do not propagate since they have no outer boundaries, or the outer boundary is very far from the well. A realistic example of an "infinite fracture" is a fracture linked to a natural fracture network that extends for kilometers. Sometimes, natural fractures are limited in size and are not linked to a fracture network. In this case, *fracture propagation* may occur if the wellbore pressure is sufficiently high. Propagation of a natural fracture is similar to the propagation of a drilling-induced fracture considered in the next section. "Growth" or "propagation" here means advancement of the fracture front whereby the rock is broken and new areas of the rock are exposed to the fluid.

As emphasized earlier, the wellbore pressure required to start or continue opening of a natural fracture depends, among other things, on the *orientation* of the fracture. Unlike induced fractures, whose orientation is determined by the least resistance principle, a natural fracture can have any orientation. Orientation of a natural fracture is generally unknown, which makes prediction of lost circulation in naturally fractured formations difficult, if possible at all. Since many natural fractures with different orientations can be exposed when drilling an interval, the concept of lost-circulation pressure in such rocks should be replaced by a distribution, or *spectrum*, of lost-circulation pressures.

5.5 LOSSES CAUSED BY INDUCED FRACTURES

Most rocks contain natural fractures, either isolated or linked to a larger fracture network. Natural fractures are potential escape paths for the drilling fluid in overbalanced drilling, as we have seen in the preceding section. However, even if there are no large natural fractures, mud losses can and do occur if fractures are created during drilling. For instance, losses into induced fractures were reportedly accountable for over 90% of lost circulation expenditures at one major operator [17]. It is commonly accepted that this happens when the BHP exceeds the so-called "fracturing pressure." There is a consensus that the fracturing pressure depends on the minimum in situ stress, and that fractures are induced along the least resistance path; ie, so that their planes are normal to the minimum in situ stress. For example, if a vertical well is drilled in a formation with $\sigma_v > \sigma_H > \sigma_h$, and the wellbore pressure exceeds the fracturing pressure, two fractures will be generated at the borehole wall. The fractures will be oriented normal to the minimum in situ stress, σ_h. If the rock is perfectly homogeneous in terms of rock mechanical properties, the fractures will initiate opposite to each other on the borehole wall. Otherwise, in a heterogeneous rock, the exact location of the fracture initiation points may be somewhat offset from the location determined by σ_h. If, on the other hand, the in situ stresses are such that $\sigma_v < \sigma_H = \sigma_h$, a horizontal fracture can be initiated from the vertical well.

In order to understand lost circulation caused by induced fractures, it is instructive to examine in detail the process of fracture initiation and propagation. A useful visualization of this process can be achieved by considering the first injection cycle in the XLOT (chapter: Stresses in Rocks).

Let us assume that the borehole wall is initially intact; ie, there are no preexisting natural fractures or flaws. If the wellbore pressure is increased to the fracture initiation pressure (FIP), a fracture will be induced at the borehole wall. The value of the FIP depends on the in situ state of stress, the borehole orientation with regard to the in situ stresses, the tensile strength of the rock, and the hydraulic conditions at the borehole wall.

Consider first an idealized case where an impermeable filter cake is rapidly deposited on the borehole wall. Under such conditions, there is no hydraulic communication between the wellbore and the formation. Consider a vertical borehole drilled in a formation with $\sigma_v > \sigma_H = \sigma_h$. In this case, a fracture will be induced when the wellbore pressure exceeds the FIP given by $(2\sigma_h - P_p + T_0)$; ie, when the effective hoop stress, σ'_θ, becomes tensile and equal to the tensile strength of the rock, T_0, cf. Eq. [2.24]. In practice, as pointed out in Ref. [23], the FIP value may be lower because the hoop stress concentration around the well can be reduced by rock plasticity.

If the horizontal in situ stresses are anisotropic—ie, $\sigma_v > \sigma_H > \sigma_h$—the FIP will be given by $(3\sigma_h - \sigma_H - P_p + T_0)$, cf. Eq. [2.38]. Hence, the stress anisotropy (σ_H being greater than σ_h) acts so as to reduce the FIP. Here, again, the borehole wall is assumed impermeable, and the pore pressure is assumed to be constant and equal to P_p everywhere behind the borehole wall.

Consider now another idealized case, where no filter cake is deposited on the borehole wall. Hydraulic communication thus exists between the borehole and the formation. If the borehole axis is parallel to the vertical in situ stress, σ_v, and the other two in situ stresses are equal, $\sigma_H = \sigma_h$, a vertical fracture will be initiated when the wellbore pressure exceeds the FIP given by [23]

$$\text{FIP} = \frac{2\sigma_h - \alpha\frac{1-2v_d}{1-v_d}P_{p0} + T_0}{2 - \alpha\frac{1-2v_d}{1-v_d}} \qquad [5.20]$$

where P_{p0} is the initial pore pressure before drilling.

A horizontal fracture may be initiated if the wellbore pressure exceeds the FIP given by [23]

$$\text{FIP} = \frac{\sigma_v - \alpha\frac{1-2v_d}{1-v_d}P_{p0} + T_0}{1 - \alpha\frac{1-2v_d}{1-v_d}} \qquad [5.21]$$

Which of the fractures, vertical or horizontal, is initiated under specific downhole conditions, will depend on the magnitudes of in situ stresses, pore pressure, and tensile strength.

An important observation about the FIP evident from the above is that FIP is a function of the in situ stresses, the pore pressure, and the degree of hydraulic communication between the well and the rock surrounding the well.

It should be emphasized at this point that initiation of an induced fracture by no means implies mud losses. Moreover, it does not automatically imply that the fracture is unstable or that it will grow further. For the fracture to start propagating, the wellbore pressure must exceed the formation breakdown pressure (FBP). FBP is defined as the wellbore pressure at which the drilling fluid enters the fracture, which leads to (unstable) fracture propagation. The value of FBP can be different from FIP because the drilling fluid generally has a sealing capacity and may thus be unable to enter the fracture that has been initiated (FIP may be equal to FBP for some types of drilling fluids; eg, pure brine without solid particles). The sealing capacity and its role in fracture reopening were discussed in Section 5.4 in connection with natural fractures. An induced fracture, unlike a natural fracture, has very small or zero initial aperture. Moreover, the aperture at the tip (the front) of an induced fracture is *always* zero. It means that even a thin drilling fluid will not be able to reach the fracture tip. As a consequence, the fracture tip is protected from the pressure buildup in the rest of the induced fracture.

Immediately upon the fracture initiation, the drilling fluid can enter only a short distance into the fracture, or not enter at all, depending on the fracture aperture, the fluid rheology, and the solids content. The fracture needs to open sufficiently wide at the mouth to enable entry of the drilling fluid. This process is schematically shown in Fig. 5.9.

As the fluid invades the fracture, and the fracture propagates, three distinct zones can be found in the fracture. In the widest part of the fracture, closer to the mouth, the fracture is filled with the fresh drilling mud coming from the well. Farther away from the mouth, the fracture aperture becomes smaller, and the fracture becomes sealed by the dehydrated drilling fluid. The aperture value that can be sealed depends on the type of the drilling fluid and is greater for a fluid with better filtration properties. Behind the seal—ie, between the dehydrated zone and the fracture tip—a noninvaded zone exists (Fig. 5.10).

The propagation of the fracture driven by the drilling fluid proceeds as follows: If the wellbore pressure is sufficiently high, it can break the seal formed by the dehydrated mud. The resistance of the seal is determined by the shear strength of the dehydrated mud, which is the reason why this parameter is used to quantify the efficiency of lost circulation and loss prevention materials. In addition to punching or shearing of the seal, the increase in the

FIGURE 5.9 Initiation of an induced fracture: (A) before initiation; (B) the fracture has opened near the mouth, but the drilling fluid has not entered the fracture; (C) the fracture has opened further, and the drilling fluid has entered part of the fracture. The fracture plane is parallel to the borehole axis in this example. The minimum in situ stress acts in the vertical direction in this Figure. Fracture aperture and length are not to scale. Drilling fluid is shown as *black fill*.

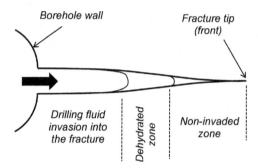

FIGURE 5.10   Three zones in an induced fracture (not to scale). The *arrow* shows the direction of fluid flow and fracture propagation. *Based on Morita N, Black AD, Fuh G-F. Borehole breakdown pressure with drilling fluids − I. Empirical results. Int J Rock Mech Min Sci Geomech Abstr 1996;33(1):39−51.*

wellbore pressure opens the fracture wider. At some point, this may destabilize the seal, and it will break. The fluid can then invade deeper into the (now wider) fracture, driving the fracture deeper into the formation. The fluid invasion and fracture propagation continue until the fluid reaches the aperture that is sufficiently small to build a new seal. The seal then builds up at the new location. The fracture propagation will stop when the wellbore pressure is equilibrated by the combined effect of fracture sealing (impeding the fluid invasion) and tensile strength of the rock (impeding the fracture growth).

The two-dimensional picture shown in Fig. 5.10 is oversimplified. In reality, the fracture will be propagating in three-dimensional space. The fracture front will advance at the weakest spots. The fracture front will therefore not necessarily remain a straight line (in the direction normal to page in Figs. 5.9 and 5.10). The faces of real fractures are rough, and the degree of roughness depends on the type of the rock [24]. These factors will result in the dehydrated zone being punched at some selected spots rather than being broken simultaneously along the entire fracture front.

Laboratory experiments with WBMs suggest that the length of the dehydrated zone increases as the fracture propagates. With OBMs, the length of the dehydrated zone remains approximately constant during the fracture growth [22]. The length means here the size of the dehydrated zone in the direction normal to the fracture front (the distance between the two leftmost dashed lines in Fig. 5.10).

The length of the noninvaded zone increases with Young's modulus of the rock (longer for stiffer rocks), decreases with in situ stress normal to fracture, and is larger with WBM than with OBM [22].

The FBP depends on the following factors [21,22]:

- in situ stresses;
- orientation of the well;

- tensile strength of the rock;
- cooling of the near-well area by the drilling fluid;
- borehole diameter;
- Young's modulus of the rock;
- permeability of the rock;
- capacity of the drilling fluid to build a filter cake and seal the fracture.

The theory developed by Morita et al. [21] enables one to estimate the FBP if detailed knowledge of the above factors is available. Unfortunately, this is rarely the case, in real life. The theory of Ref. [21] provides, however, valuable insights into the effect of different operational and in situ factors on mud losses caused by induced fracturing. It was the first theory of lost circulation that offered a consistent description of induced fracturing and emphasized the crucial role of fracture tip isolation from the rest of the fracture. Today, this theory serves as a basis for design and application of lost circulation and loss prevention materials (chapters: Preventing Lost Circulation and Curing the Losses).

Tensile strength of the rock has an obvious effect on the FBP. In practice, the tensile strength of the rock is often set equal to zero in calculations because rocks usually contain small preexisting cracks. If such crack intersects the borehole wall, they will effectively reduce the strength of the rock to zero at their locations. However, a small natural crack is not necessarily located at the most likely induced-fracture initiation point. The rock may thus very well have nonzero tensile strength. The tensile strength of many sedimentary rocks is, however, so low compared to the in situ stresses that its contribution to the FBP is negligible.

Rock cooling by the drilling fluid reduces the hoop stress and thereby reduces both the FIP and the FBP.

Larger borehole diameter and smaller Young's modulus of the rock make it easier to open the fracture, thereby reducing the FBP.

Filtration properties have a major impact on the ability of the fluid to seal the induced fracture. These properties encompass filtration properties of both the drilling fluid and the rock. In particular, laboratory experiments conducted on Mancos shale and Berea sandstone demonstrated the following [22]:

1. Low-permeability rocks have smaller FBP than high-permeability rocks.
2. In high-permeability specimens with a prefractured borehole, injecting a WBM produced a higher FBP than injecting OBM did. On the other hand, when the borehole wall was initially intact (no prefracturing), there was as good as no effect of the mud type on the FBP value.
3. Fracture propagation is more unstable in low-permeability rocks. In such rocks, the injected fluid has nowhere else to go but into the induced fracture. In high-permeability rocks, on the other hand, the fluid can drain through the fracture wall into the rock. Thereby, less fluid remains available to extend the fracture.

The experiments in Ref. [22] were performed under controlled fluid injection rate. Such regime is more typical of hydraulic fracturing or an XLOT rather than a lost circulation event. For this reason, some of these experimental findings should be applied for lost circulation with caution. However, notwithstanding the difference between the laboratory and in situ boundary conditions, more unstable and less controlled fracture propagation should indeed be expected during lost circulation in low-permeability rocks (eg, shale) than in high-permeability rocks (eg, sandstone). Also, the effect of filtration properties on the FBP observed with prefractured boreholes should be fairly insensitive to the difference between the injection regimes in the lab and in the field.

The question about the "correct" type of boundary conditions at the fracture mouth to be used in laboratory tests is not an easy one. This question is touched upon in the next section, where we discuss the interplay between the well flow and the fracture flow during a lost circulation event.

The discussion so far in this section suggests that, in the case of an intact circular borehole wall, the FBP is greater than the FIP. This holds true as long as there are solids in the drilling fluid that can be deposited in the fracture, isolating the fluid pressure in the noninvaded zone from the rest of the fracture. Such solids are present in WBM, OBM, and synthetic-base mud (SBM). If no such solids are available, or their sealing capacity is poor (eg, when drilling with clear water), the values of FBP and FIP could be expected to be close or equal. Laboratory experiments by Morita et al. [22] revealed, however, that the FBP value is higher than the theoretical FIP value even with solids-free fluids. Also with such fluids, a stable fracture can be initiated when the borehole pressure exceeds FIP, provided that the rock has sufficiently high permeability. Morita et al. [22] attributed the stable fracture propagation in this case to the leakoff and high frictional pressure losses inside the thin initial fracture, which helps stabilize the fracture propagation.

So far, we have looked at the fracture initiation and formation breakdown. The question now is: what happens *after* the formation breakdown?

The borehole pressure required to extend the fracture outside the near-well area is the fracture propagation pressure (FPP). It is given by a sum of three components: (i) the pressure required to create new fracture surfaces (which is a function of rock's tensile strength); (ii) the pressure required to overcome the pressure loss inside the fracture, from the borehole to the fracture tip; (iii) the pressure required to counterbalance the in situ stress normal to the fracture plane, thereby keeping the fracture opened. The FPP will be somewhat higher than the minimum in situ stress because of contributions (i) and (ii).

Outside the near-well stress concentration zone, the fracture will assume the most favorable orientation; ie, such that it requires least energy for propagation. The fracture plane will, thus, become normal to the minimum in situ stress. Let us consider a few simple examples.

The first example is a vertical well drilled in a formation with $\sigma_v > \sigma_H > \sigma_h$. If the wellbore pressure increases above the FIP, a vertical fracture will be initiated, oriented perpendicular to the minimum in situ stress, σ_h. The fracture will remain normal to σ_h as it propagates, unless it is deflected by rock heterogeneities (for instance, hard inclusions).

As a second example, consider an inclined or a horizontal well drilled in a formation with $\sigma_v > \sigma_H > \sigma_h$. An induced fracture might initially not be perpendicular to the far-field minimum in situ stress, σ_h, since the direction of the minimum stress in the near-well area is not necessarily parallel to σ_h. As the fracture front leaves the near-well stress concentration zone, the fracture will gradually turn so as to become normal to σ_h.

As a third example, consider a vertical well drilled in a formation with isotropic horizontal in situ stresses, $\sigma_v > \sigma_H = \sigma_h$. In this case, the induced fracture will be vertical, but there is no preferential azimuthal orientation of the fracture. The fracture will be initiated at the azimuthal position on the borehole wall where the rock has the lowest strength.

As our final example, consider a vertical well drilled in a formation with $\sigma_v < \sigma_h < \sigma_H$. The induced fracture will most likely be horizontal in this case.

If the drilling fluid is not capable of sealing the fracture (eg, drilling with clear water), the pressure loss inside the fracture (component (ii) in the FPP) is solely due to the pressure losses caused by friction between the fluid and the fracture walls. However, solids-laden fluids, such as WBM or OBM, create a seal (a dehydrated zone) between the fracture tip and the part of the fracture opened to the well (Fig. 5.10). The fluid then needs to break through the seal in order to move the fracture tip deeper into the formation. The breakthrough occurs at the weakest locations [24]. Thus, with drilling fluids having a sealing capacity (WBM, OBM, SBM), component (ii) is increased by the amount needed to break through the seal. In addition, the seal protects (at least to some extent) the noninvaded zone from the fluid pressure in the rest of the fracture. Thus, the fluid pressure near the tip is smaller than it would be if the fluid were not sealing. This increases the FPP further.

During fracture propagation, the seal is periodically built up and broken. The pressure at the fracture mouth will therefore fluctuate around the average FPP value. The pressure slightly rises as the seal builds up, and then slightly drops as the seal is broken at one or several locations. The magnitude of the pressure fluctuations depends on the filtration capacity of the drilling fluid. The fluctuations are larger with the mud that has higher filtration through the fracture wall [22]. In general, therefore, the pressure fluctuations during fracture propagation will be greater with WBM than with OBM. In addition to the pressure fluctuations caused by the seal buildup and breakage, some variation in the FPP may be caused by rock heterogeneity ($T_0 \neq$ const) and fluctuations in the fracture trajectory (the fracture surface is not a perfect plane).

The FPP depends on the following factors [25]:

- minimum in situ stress;
- tensile strength of the rock;
- injection flow rate;
- permeability of the rock;
- capacity of the drilling fluid to build a filter cake and seal the fracture.

As the fracture propagates away from the well, the increasing frictional pressure loss between the well and the fracture tip tends to increase the FPP value. On the other hand, increasing fluid-invaded fracture length and fracture aperture at the mouth tend to reduce the FPP. In addition, the fracture front becomes longer as the fracture extends. As pointed out by Økland et al. [25], this will increase the probability of having a weak spot at some location along the edge, which might reduce the FPP.

The fracture propagation process outlined above can be complicated by heterogeneities in rock properties and in situ stresses. Furthermore, preexisting (natural or depletion-induced) fractures found in most rocks may deflect or arrest the induced fracture [26].

Based on the discussion so far, the lower bound of FPP can be assumed to be the sum of the minimum in situ stress and the tensile strength of the rock. In an extensional tectonic environment ($\sigma_v > \sigma_H > \sigma_h$), this yields:

$$\text{FPP} \geq \sigma_h + T_0 \qquad\qquad [5.22]$$

The value on the right-hand side of Eq. [5.22] may be smaller or larger than the FIP and FBP. Consider, for instance, a vertical well drilled in a formation with $\sigma_v > \sigma_H > \sigma_h$. If perfect impermeable filter cake builds up on the borehole wall, the FIP is given by ($3\sigma_h - \sigma_H - P_p + T_0$). If σ_H is much higher than σ_h, the FIP (and possibly FBP) will be lower than ($\sigma_h + T_0$). If σ_H is close to σ_h, FIP (and possible FBP) will be greater than ($\sigma_h + T_0$). In any event, it is the value of FPP that controls the fracture extension beyond the near-well area and thus determines the severity of mud losses. The relative magnitude of FBP affects, however, whether the fracture leaves the near-well area (FBP > FPP, making the fracture propagation after the breakdown unstable) or is arrested near the wellbore (which might happen if FBP < FPP and BHP < FPP). In the latter case, minor losses or borehole ballooning (reversible losses/gains discussed further in this chapter) may be observed. They do not, however, represent a serious lost-circulation problem [27].

Other factors affecting the extension of the fracture are the leakoff through the fracture wall, and the pumping conditions. Consider, for example, a hypothetical case where fluid is pumped into an induced fracture with a constant flow rate. If the rock is impermeable (such as, in practice, some shales are), all the fluid pumped into the fracture is used to extend it. In this case, the fracture will grow indefinitely, and the relation between FBP and FPP (whether FBP > FPP or vice versa) does not matter. If the rock is permeable, and sufficient leakoff occurs through the fracture wall, the leakoff will reduce the fracture propagation speed and possibly bring the fracture to a halt.

Consider now a different hypothetical case in which the fluid pressure at the fracture mouth is fixed. This corresponds to the BHP remaining constant during a lost circulation event, again a thought scenario only. In this case, the relation between FBP and FPP does matter. In particular, if FBP < FPP and the BHP is such that FBP < BHP < FPP, the fracture will stop before leaving the near-well area. If the BHP exceeds FPP at the early stages of propagation, the fracture will grow outside the near-well stress concentration area until the FPP becomes greater than the applied BHP (due to, eg, frictional pressure losses in the long fracture).

If FBP > FPP, there will be unstable fracture propagation after the breakdown. This should not normally happen since, as we will see later in this chapter, the upper operational bound of the BHP is usually set close to the minimum in situ stress. If the BHP is below FBP, there will be no induced fracturing as long as the borehole wall is circular and intact; ie, does not contain any cracks, flaws, etc. that might act as fracture nuclei.

In reality, the FBP value can be lowered in the presence of initial cracks. The location, orientation, and size of such cracks are not known before they are intersected by the well. Therefore, it is difficult to ensure that there will be no losses as long as the BHP remains below the FBP calculated from the hoop stress concentration. Instead, the value of σ_h is often chosen as the upper bound of the operational BHP. As long as BHP < σ_h, the induced fracture will not be able to propagate outside the near-well area. Hence, if induced fracturing occurs, mud losses will be limited by the volume of the relatively short fractures (plus whatever leakoff takes place through the fracture wall until filter cake builds up or the fluid pressure equilibrates between the fracture and the rock around it).

In addition to the pressure values such as FIP, FBP, and FPP, the process of induced fracturing outlined in this section is characterized by two additional parameters. These are:

- the minimum fracture aperture that can be entered by fresh mud from the borehole, w_{c1};
- the maximum fracture aperture that can be sealed by dehydrated mud inside the fracture, w_m.

We will see in chapters "Preventing Lost Circulation" and "Curing the Losses" that these two parameters play important role when designing preventive and remedial treatments of lost circulation. In particular, a drilling fluid that has higher w_m—ie, can seal a larger aperture—gives rise to higher FPP [21]. The values of w_m on the order of 0.3 mm are obtained for typical drilling fluids [21].

5.6 SEVERE LOSSES, TOTAL LOSSES, AND BOREHOLE BALLOONING

It was mentioned in the previous section that laboratory studies of mud losses and induced fracturing are usually performed with constant injection rate into the fracture. This could represent a field scenario where a section of an open hole is isolated (eg, by packers), and fluid is injected with constant flow rate into this isolated interval. While such conditions may indeed exist—eg, in a minifrac test—mud losses during drilling occur while there is flow both into and out of the well. It is therefore not evident that constant flow rate into the fracture always is the best choice of the boundary condition for the fracture. The inadequacy of the constant flow rate assumption was recognized by Adachi et al. who employed constant wellbore pressure as the boundary condition in their theoretical model of mud losses into a natural fracture [28]. It is however, not obvious that such boundary conditions is correct either.

In reality, a fracture (natural or induced) or a permeable zone is coupled to the flow in the well. The issue of boundary conditions at the fracture mouth requires therefore analysis of the coupled well/fracture flow. In order to illustrate the problem, we construct a simple model in this section that will also illuminate the difference between severe mud losses and total loss of returns.

Consider a vertical onshore well. Assume that the well is drilled with an incompressible Newtonian fluid (clear water), and the drilling fluid has the same properties (viscosity, density) as the formation fluid. At depth D the well meets a preexisting, natural fracture. We assume that the fracture is horizontal, has circular shape, and the well punches the fracture in the center. We also assume that the vertical in situ stress is sufficiently high so that the BHP never exceeds the FRP during the mud loss incident. The fracture is assumed to be linked to an aquifer (or a large fracture system) at its far edge, so that the fluid pressure at the edge remains constant and equal to the original formation pore pressure, P_{p0}, at all times. Under these simplifying assumptions, the model is a toy model indeed. However, the purpose of our analysis here is only to highlight the most essential features of a mud loss event. With this purpose in mind, the above simplifications are acceptable.

The geometry of the problem is illustrated in Fig. 5.11. The well is assumed to have radius R_w along its entire depth, D. The drillpipe is assumed to have the outer radius, R_d; ie, there are no constrictions in the annulus (drill collars, etc.). These assumptions simplify the calculation of the annular flow without any consequences for the general results we seek.

The mass balance in the system is given by:

$$Q_{an} = Q_{pump} - Q_{frac} \qquad [5.23]$$

where Q_{an} is the flow rate in the annulus (positive upwards); Q_{pump} is the pump rate; Q_{frac} is the flow rate from the wellbore into the fracture (positive: mud

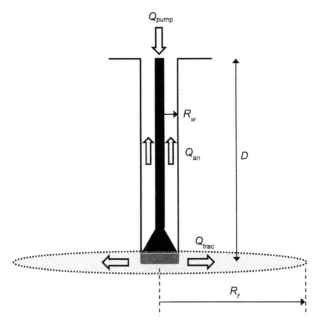

FIGURE 5.11 Geometry of the model of a mud loss event.

loss; negative: mud gain). The pump rate is kept constant during the mud loss event in our analyses.

The momentum balance in the annulus is given by:

$$P_w = \Delta P_{static} + \Delta P_{APL} \qquad [5.24]$$

where P_w is the BHP (in excess of the atmospheric pressure)—ie, the fluid pressure at the fracture mouth; ΔP_{static} is the hydrostatic pressure of the mud column in the annulus; ΔP_{APL} is the annular pressure loss (APL). It is assumed in Eq. [5.24] that the annulus is opened to the atmosphere; ie, there is no backpressure on the return line.

The static pressure is given by:

$$\Delta P_{static} = \rho g H \qquad [5.25]$$

where ρ is the mud density; H is the height of the mud column in the annulus. As long as there are returns, H is equal to the wellbore depth, D. Total loss of returns occurs when H becomes smaller than D.

The flow regime in the annulus is assumed to be laminar. The APL is therefore given by [29]:

$$\Delta P_{APL} = \frac{8\mu Q_{an} H}{\pi \left[R_w^4 - R_d^4 - \left(R_w^2 - R_d^2 \right)^2 / \ln(R_w/R_d) \right]} \qquad [5.26]$$

Since we only are interested in the first few minutes of the mud loss event, penetration of the drill bit beneath the fracture during the event can be neglected.

The pressure at the fracture mouth given by Eq. [5.24] can, finally, be used to calculate the fracture flow rate, Q_{frac}, to be used in Eq. [5.23]. Assuming that the fracture has a constant hydraulic aperture, w_h (BHP remains below FRP, by assumption), the fracture flow rate is given by:

$$Q_{\text{frac}} = \frac{\pi w_h^3 (P_w - P_p)}{6\mu \ln(R_f/R_w)} \qquad [5.27]$$

Once the annular flow rate is obtained from Eq. [5.23], the change in the mud column during the simulation time step, Δt, can be evaluated. If the mud column occupies the entire annulus ($L = D$), and the annular flow is upwards, the annular mud column height remains unchanged and equal to D. If the mud column occupies the entire annulus ($L = D$), and the annular flow is downwards, the annular mud column height starts dropping. If the mud column occupies only the lower section of the annulus, and thus there are no returns, the annular mud column height increases if $Q_{\text{an}} > 0$ and decreases if $Q_{\text{an}} < 0$.

To cover all the above scenarios, the change in the mud column height during time step Δt is given by:

$$\Delta H = \frac{\Delta H}{\Delta t} \cdot \Delta t \qquad [5.28]$$

with

$$\frac{\Delta H}{\Delta t} = \begin{cases} 0, & \text{if} \quad Q_{\text{an}} > 0 \text{ and } H = D \\ Q_{\text{an}}/A_{\text{an}}, & \text{if} \quad Q_{\text{an}} > 0 \text{ and } H < D \\ Q_{\text{an}}/A_{\text{an}}, & \text{if} \quad Q_{\text{an}} < 0 \end{cases} \qquad [5.29]$$

where A_{an} is the area of the annular cross-section normal to flow, $A_{\text{an}} = \pi(R_w^2 - R_d^2)$.

The initial conditions for the model are as follows: The fracture flow rate is initially zero (no flow in the fracture before it is hit by the drill bit):

$$Q_{\text{frac}} = 0 \text{ at } t = 0 \qquad [5.30]$$

The annulus is initially filled with mud; ie, drilling is with full returns until the drill bit hits the fracture:

$$H = D \text{ at } t = 0 \qquad [5.31]$$

We now consider two examples. All the parameters, except the pump rate, are identical in the two examples and are given in Table 5.2. The pore pressure is equal to $0.99\rho g D$; ie, subnormal.

TABLE 5.2 Model Parameters Used in Examples 1 and 2

Parameter	Value
Depth, D	1524 m (5000 ft)
Density of drilling fluid and formation fluid, ρ	1000 kg/m^3
Viscosity of drilling fluid and formation fluid, μ	0.001 Pa s (1 cP)
Radius of the drillpipe, R_d	0.127 m (5 in.)
Radius of the well, R_w	0.178 m (7 in.)
Radius of the fracture	100 m
Hydraulic aperture of the fracture, w_h	1 mm
Pore pressure, P_p	14.7858 MPa

Example 1. Pump rate $Q_{pump} = 10$ L/s.

The formation is slightly underpressured. Hence, drilling fluid would be lost into the fracture even if the pumps were off. In our example, the pump is on. The fluid is Newtonian, therefore it will always be able to enter and flow in the fracture (sealing capacity of the fluid is zero, and it has no yield stress). The calculated flow rate into the fracture as a function of time is shown in Fig. 5.12A. The drill bit hits the fracture at time $t = 0$. The flow rate into the fracture increases suddenly from 0 to 12.5 L/s at the beginning of the event, and then gradually decreases to 10 L/s, which is equal to the pump rate. It means that all the fluid pumped into the well escapes into the fracture. The mud in the annulus starts falling at $t = 0$ (Fig. 5.12B). The mud level stabilizes at 1521.1 m, which is the head needed to balance the formation pressure plus the pressure drop in the fracture. After just a few minutes, a dynamic equilibrium is established whereby all the fluid pumped into the well goes into the fracture, while the BHP remains constant and equal to the static pressure of the annular mud column (Fig. 5.12C). This example illustrates *total losses*.

In reality, the drilling fluid is usually a non-Newtonian yield-stress fluid. Therefore, the fluid may stop flowing in the fracture at some time, as explained earlier in this chapter. If, however, the fracture is sufficiently wide (eg, fractures encountered in geothermal drilling), the scenario described in this example becomes quite realistic. In addition, we neglected the fracture reopening in this example, assuming that the in situ vertical stress is so high that the fracture stays mechanically closed at all time. In reality, the fracture or parts of it may reopen, which will further increase losses.

Example 2. Pump rate $Q_{pump} = 13$ L/s.

In this example, the pump rate of 13 L/s is so high that the fracture cannot accept all the fluid pumped down the well. Hence, some of the fluid will return

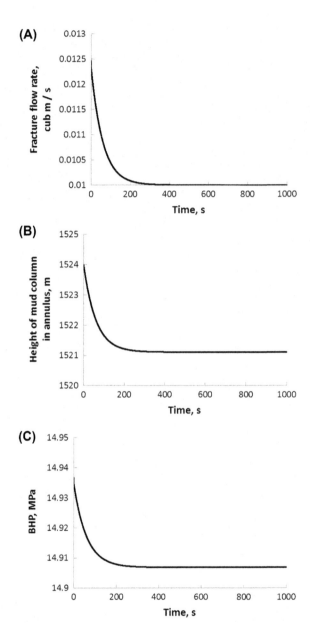

FIGURE 5.12   Simulation of a mud loss event. The drill bit hits a natural fracture at time $t = 0$. Pump rate is constant and equal to 10 L/s. (A) mud flow rate into the fracture as a function of time; (B) height of the mud column in the annulus as a function of time; (C) bottomhole pressure as a function of time.

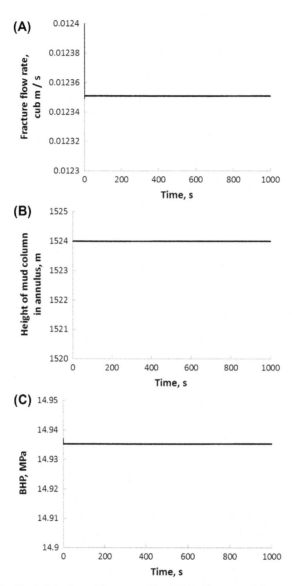

FIGURE 5.13 Simulation of a mud loss event. The drill bit hits a natural fracture at time $t = 0$. Pump rate is constant and equal to 13 L/s. (A) mud flow rate into the fracture as a function of time; (B) height of the mud column in the annulus as a function of time; (C) bottomhole pressure as a function of time.

to surface. The final flow rate into the fracture is 12.35 L/s; ie, higher than in Example 1. The equilibrium is established almost instantaneously after the drill bit hits the fracture (Fig. 5.13). The mud column in the annulus does not drop, drilling is with partial returns. Whether these partial returns are sufficient to transport drill cuttings and keep the hole clean depends on many factors, including the rate of penetration. The mud loss flow rate of 12.35 L/s indicates severe losses.

Sudden onset of mud losses observed in Examples 1 and 2 is typical of losses caused by natural fractures (Fig. 5.3).

It was mentioned earlier in this chapter that laboratory studies of lost circulation are usually performed with such boundary conditions as constant flow rate into the fracture or constant pressure at the fracture mouth. Figs. 5.12 and 5.13 suggest that such boundary conditions are not always adequate. Indeed, even in our oversimplified model, neither the flow rate into the fracture nor the pressure at the fracture mouth remained constant during the event. In real life, the situation could be further complicated by different fluid rheology in the fracture and in the well, possible fracture reopening, etc. Constant flow rate or constant BHP can therefore only be considered as idealized boundary conditions that are convenient in laboratory experiments or numerical models, but cannot properly describe the real downhole conditions.

Fluid lost into the fracture in Examples 1 and 2, is lost forever. It cannot be recovered even if the pump is turned off. Quite a different situation is when mud losses are due to a short, isolated fracture. If the BHP exceeds the FRP but remains below the FPP, the fracture will open and will accept mud. If the BHP is then reduced, the fracture will close, and some of the mud will flow back into the well. This phenomenon is known as *borehole ballooning* or *borehole breathing*. It can be defined as reversible (or partially reversible) loss/gain of the drilling fluid as the BHP increases/decreases. The qualifier "partial" is here to emphasize that it may be difficult to squeeze all of the fluid that entered the fracture, back into the well. The higher the yield stress of the mud, the more difficult it is to enter the fracture, but also the more difficult it will be to flow back after the BHP is reduced. The volume of losses and gains during ballooning may vary and typically is in the range of $25-350$ bbl ($4-55$ m^3) [27].

In addition to fracture opening/closing, ballooning can be caused by compressibility of the fluid in the well, by thermal fracturing due to the cooling effect in the near-well area, or by thermal expansion/contraction of the drilling fluid in the annulus. The latter may happen, for example, when circulation is stopped to make connection. During circulation, relatively cold drilling fluid is pumped down the drillpipe. The fluid is heated as it travels down the pipe, but typically is still colder than the surrounding formation as it exits through the drill bit nozzles. As the fluid travels up the annulus, it is colder than the rock in the lower part of the annulus. At some depth, the temperature of the mud is equal to the temperature of the formation. The mud is warmer than the formation (or the casing pipe) as it travels further up the annulus and exits at the surface. This is schematically illustrated in Fig. 5.14. The temperature of the mud during circulation is thus nowhere equal to the temperature of the surrounding rocks, except at one location. Therefore, when the circulation is stopped, the mud temperature in the annulus will change so as to establish a thermal equilibrium with the formation. The result can be some mud gain at

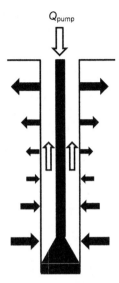

FIGURE 5.14   Thermal flux from and to the well during circulation (*solid black arrows*). *White arrows* show the fluid flow direction.

the pits [30]. The equilibrium is reached when the temperature profile in the well follows the geothermal temperature profile. The volume of losses and gains caused by the thermal mechanism may amount to several percent of the wellbore volume, and is greater in the case of OBM than WBM [30].

Ballooning is quite common, for example, in naturally fractured rocks. According to Ameen [31], 32% of wells in the Khuff Formation (Saudi Arabia) experience borehole ballooning.

Mud gains observed as part of a ballooning event may be misinterpreted as the onset of a kick. Therefore, it is important to have robust procedures for discriminating between ballooning and lost circulation (or a kick). Numerical models, such as the one developed in Ref. [32], may help in establishing the "flow rate versus time" patterns characteristic of ballooning.

5.7 WHAT SHOULD WE CHOOSE AS THE UPPER BOUND OF THE OPERATIONAL PRESSURE WINDOW?

Several mechanisms of lost circulation have been discussed in this chapter. Even the most sketchy analysis of these mechanisms reveals a great variety of lost circulation scenarios. As a consequence, we have to deal with a conundrum of different estimates for the maximum operational BHP. Choosing this value correctly is of utmost importance for cost-efficient drilling. Setting the upper operational bound of BHP too low may prevent lost circulation, but will increase the number of casing points. Setting the upper operational bound too high will result in mud losses, with nonproductive time and extra costs as a consequence.

As we have seen, there is no unique equation that would predict the lost-circulation pressure. One of the reasons for the lack of such magic equation is that there are several different mechanisms at play. For instance, losses into highly permeable or vugular formations occur when the BHP exceeds the formation pore pressure. In situ stresses play virtually no role here. On the contrary, lost circulation caused by induced fractures occurs at the BHP that depends on the in situ stresses. In the presence of preexisting natural fractures, in situ stresses affect the lost circulation, and so does the orientation of the fractures. Unlike induced fractures, natural fractures may have any orientation with regard to the borehole, and the lost-circulation pressure is thus unknown before drilling since there is no way to know the orientation of a specific fracture before it is actually exposed at the drill bit.

What is needed in practice is a number, the upper operational bound of the BHP that would enable drilling without lost circulation. Since no such unique number can be produced, and since too many unknowns are involved (location, aperture and orientation of natural fractures, heterogeneity of rock strength, etc.), it seems logical to set the upper bound of the BHP so as to guarantee that losses, if they occur, remain manageable.

In the case of induced fractures, this would imply that the fracture remains relatively short; ie, does not propagate outside the near-well area. To achieve this, it is sufficient to require that the BHP stay below the minimum in situ stress. As we have seen, the FIP and FBP can be either greater or smaller than the minimum in situ stress. Hence, setting the BHP equal to or slightly lower than the minimum in situ stress might not prevent fracture initiation. It will, however, prevent uncontrolled fracture growth outside the near-well area. It is this uncontrolled fracture growth that leads to severe losses. Thus, setting the BHP equal to the minimum in situ stress will prevent severe losses caused by induced fracturing.

In the case of natural fractures, both the fracture aperture and the fracture orientation are important. Wide-opened fractures may cause severe losses as soon as the BHP exceeds the fluid pressure inside the fracture (and especially if the fracture is long and connected to a fracture network). Such situations are common in geothermal drilling. On the other hand, if the hydraulic aperture of a natural fracture is sufficiently small, losses might stop due to yield-stress rheology of the drilling fluid. In this case, to induce severe losses, the BHP should be high enough to reopen the fracture *and* maintain flow (ie, overcome the pressure losses and leakoff inside the fracture). The reopening does not occur until the BHP exceeds the in situ stress normal to the fracture plane. The worst-case scenario is when the fracture plane is perpendicular to the *minimum* in situ stress. Therefore, the minimum in situ stress can be chosen as the upper operational bound of the BHP in this case. However, in some cases (wide-opened natural fractures), lost circulation might still occur even though the BHP is below this value. In other cases, lost circulation will not occur even when the BHP exceeds this upper bound, particularly if the fluid is thick and thus has high frictional losses inside the fracture, or if the fracture plane is not perpendicular to the minimum in situ stress.

The case of highly-porous or vugular formations is somewhat similar to the wide-opened fractures. The losses start as soon the BHP exceeds the pore pressure. The onset of lost circulation is therefore not affected by the in situ stresses. Since the pore pressure in a rock is lower than the minimum in situ stress, setting the upper operational bound of BHP equal to the minimum in situ stress cannot prevent losses in this case. In fact, losses cannot be prevented as long as the BHP is above the pore pressure, which is always the case in overbalanced drilling. This is the reason why vugular formations, high-permeability rocks and wide-open fractures represent the most difficult cases. Lost circulation in such formations can be reduced or eliminated if UBD or managed pressure drilling (MPD) is employed. In particular, in the case of MPD, the BHP can be maintained at or close to the formation pore pressure, thereby avoiding both mud losses and kicks. Basics of MPD and UBD are reviewed in chapter "Preventing Lost Circulation," along with other preventive measures.

The above arguments suggest that setting the upper operational bound of BHP equal to the minimum in situ stress will prevent lost circulation in many cases. This is the reason why the minimum in situ stress is often chosen as the upper operational bound of BHP in overbalanced drilling. The minimum in situ stress is usually obtained by stress measurements; eg, XLOT (chapter: Stresses in Rocks). A safety margin is usually applied; ie, the upper operational pressure is set slightly lower than the minimum in situ stress. This approach, however, might not prevent losses in vugular or high-permeability formations as well as in the presence of long, wide-opened fractures. Modifications to drilling procedures such as MPD may be required in order to drill such formations. In some other cases, the minimum in situ stress provides a conservative estimate, and the well could in fact be drilled with a higher BHP.

5.8 LOST CIRCULATION IN DEPLETED RESERVOIRS

Lost circulation is exacerbated in certain types of wells and formations. In this and subsequent sections, we will take a closer look at the mechanisms that are at play in such wells. We start with drilling in depleted reservoirs.

Increased lost-circulation risk in depleted reservoirs is due to the following reasons [33]:

- reduced pore pressure;
- reduced total stresses;
- depletion-induced fractures.

Pore pressure decreases during reservoir depletion. This means that using the same mud weight in the depleted formation as in the rest of the well will effectively create higher overbalance in the former. Higher overbalance will promote losses into the depleted high-permeability reservoir rock. It will also facilitate fluid flow into natural fractures.

We saw in chapter "Stresses in Rocks" that total stresses (both vertical and horizontal) decrease during depletion. In terms of lost circulation, it means that the fracturing pressure decreases. Therefore, it is easier to induce fractures in a depleted field. If the stress regime is extensional ($\sigma_v > \sigma_H = \sigma_h$ or $\sigma_v > \sigma_H > \sigma_h$), vertical fractures are more likely to be induced when drilling a depleted reservoir than when drilling the same reservoir before depletion. In a compressional stress regime ($\sigma_v < \sigma_H = \sigma_h$ or $\sigma_v < \sigma_h < \sigma_H$), initiation and propagation of horizontal fractures is facilitated by depletion.

Stress changes during depletion may result in fracturing and fault reactivation during production [34]. In particular, normal faults may be reactivated within the reservoir in extensional stress regime, and vertical fractures may be generated in the sideburden (chapter: Stresses in Rocks). Fault reactivation is usually accompanied with fracture development around faults (chapter: Natural Fractures in Rocks). These depletion-induced fractures will exacerbate lost-circulation problems when subsequently drilling such formations.

An additional complication in depleted reservoirs is due to heterogeneous depletion [35]. Pressure decrease due to depletion may vary across the reservoir. For instance, lower-permeability rocks may experience smaller pressure decline than adjacent high-permeability rocks (Box 5.3). This will result in a complex distribution of the pore pressure and fracturing pressure along the well. This is schematically illustrated in Fig. 5.15. The pore pressure and fracturing pressure profiles along a vertical well are shown as dashed lines before the depletion and as solid lines after the depletion. In the case of a homogeneous depletion (Fig. 5.15A), the entire vertical interval can be drilled

---

BOX 5.3 Mud Losses in a Depleted Reservoir in the Gulf of Mexico

A case history of mud losses in a deepwater well demonstrating devastating consequences of heterogeneous depletion was described by Willson et al. in their SPE paper 84266 "Wellbore stability challenges in the deep water, Gulf of Mexico: case history examples from the Pompano Field."

The well with the target depth of ca. 12,800 ft (3901 m) measured depth (MD) was drilled through a formation composed of depleted sands and overpressured shales. The fracture gradient in the sands was estimated at 12.45 ppg. The minimum mud weight required to maintain borehole stability in shales was estimated at 13 ppg. Following these estimates, the well was drilled with 12-1/4" diameter and 13.1 ppg mud weight down to 11,280 ft (3438 m) MD. This enabled the operator to successfully prevent influx from overpressured shales at 9777 ft (2980 m) MD, but induced total losses in sands at 11,280 ft (3438 m) MD. Eventually, as a result of this lost circulation incident, the bottomhole assembly was lost.

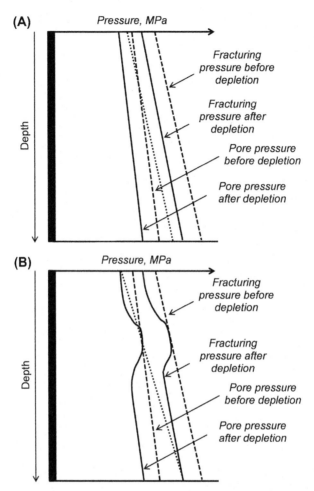

FIGURE 5.15   Pore pressure and fracturing pressure versus depth before (*dashed lines*) and after (*solid lines*) depletion. (A) Homogeneous depletion of the reservoir. (B) Heterogeneous depletion of the reservoir. *Dotted line* indicates the static BHP when drilling a vertical well.

without setting extra casing strings because the static BHP can be kept within the drilling margin: the dotted line stays within the BHP window along the entire interval, with some safety margins on both sides.

If, however, depletion is heterogeneous, it may become impossible to stay within the drilling margin with the same mud weight along the entire interval. This is illustrated in Fig. 5.15B where the BHP is forced below the pore pressure in the undepleted part of the interval if the entire interval is to be drilled with the same mud weight. An attempt to avoid drilling with BHP below the pore pressure by increasing the mud weight will lead to mud losses in the depleted parts of the interval. Conventional solution to this problem is to set an additional casing point. Alternatively, MPD can be employed.

Mature fields with water injection may have horizontal stresses additionally reduced by the cooling effect [36]. As we saw in chapter "Stresses in Rocks," the cooling-induced reduction of the total horizontal stresses in the reservoir is proportional to the temperature change. Thus, the cooling effect is most significant in the vicinity of injectors. Infill drilling in such areas may result in severe losses. As pointed out by Hettema et al. [36], the cooling effect can be so strong that the operational pressure window closes completely.

5.9 LOST CIRCULATION AND FORMATION HETEROGENEITIES

In the previous section, we saw how heterogeneous pore pressure and fracturing pressure distributions exacerbate losses in depleted formations. This is only one example of heterogeneity. In general, heterogeneous distributions of the in situ stresses or of the rock strength may exist in a reservoir even before production.

For instance, shales often exhibit higher pore pressure and higher horizontal stress than the adjacent sandstones. In addition, boundaries between shales and sandstones can act as weakness planes and escape paths for the drilling fluid. For this reasons, losses can occur when exiting shale stringers [35].

Rubble zones found at the bottom of salt bodies are notoriously difficult to drill because of severe losses. Being aware of the presence and location of such heterogeneities, and their possible detrimental effects, can help in preventing and curing lost circulation.

5.10 LOST CIRCULATION IN DEEPWATER DRILLING

Lost circulation is exacerbated in deepwater drilling due to the following reasons:

- narrow operational BHP window;
- low rock strength in shallow sediments;
- the use of a riser.

Relatively low degree of compaction is typical of overburden sediments in deepwater environments. The consequences of rapid deposition and undercompaction are as follows [37]:

- overpressure (pore pressure in excess of the hydrostatic);
- reduced density of the sediments;
- reduced strength;
- reduced vertical and horizontal in situ stresses, as a result of reduced density.

The overpressure and low horizontal stresses may significantly reduce the operational BHP window in deepwater wells. The window can be further narrowed when drilling deviated and horizontal sections in deep water (see Section 5.12).

In addition to the narrow operational BHP window, lost circulation problems in deepwater drilling become aggravated when using a riser (Fig. 5.16). It is assumed in Fig. 5.16 that shallow sediments are under normal pore pressure regime. Thus, the solid line below the sea floor is a continuation of the one above. This assumption simplifies the drawing but does not affect the conclusions. The fracturing pressure in the loose sediment on the sea floor is near the hydrostatic pressure. Thus, the solid line and the dashed line intersect at the sea floor.

Overbalanced drilling of the shallow sediments with a riser will result in the static BHP profile shown by the dotted line in Fig. 5.16. It is evident that keeping the BHP above the pore pressure in the lower part of the open hole will bring the annular pressure above the fracturing pressure higher up the hole. This will induce losses in the upper part of the hole [38].

Fig. 5.16 reveals that the cause of lost circulation in this case is the weight of the mud column in the riser. In overbalanced drilling, the static pressure in the riser at the mud line is higher than the hydrostatic sea water pressure. Hence, the BHP in the upper part of the hole is higher than the fracturing pressure.

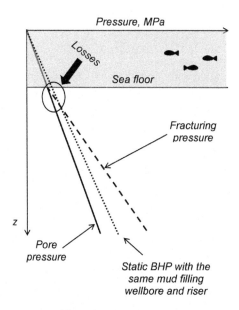

FIGURE 5.16   Pore pressure and fracturing pressure versus depth in marine drilling with riser. Lost circulation zone is encircled.

A solution to this problem known as "pump and dump" technique involves releasing the returning mud at the sea floor. A more environmentally friendly option makes use of a subsea pump to lift the mud from the sea floor to the rig. This option is one of the MPD techniques discussed in chapter "Preventing Lost Circulation."

Fig. 5.16 illustrates the cause of lost circulation in shallow formations in marine drilling. Let us now work out the theory in more detail. Consider again a vertical well drilled with a riser, as in Fig. 5.16, and let us evaluate the operational pressure window. According to Fig. 5.16, the horizontal in situ stress is given by

$$\sigma_h = \rho_w g D_{\text{SW}} + \frac{d\sigma_h}{dz} D_{\text{F}} \qquad [5.32]$$

where ρ_w is the sea water density; D_{SW} is the water depth; z is the vertical coordinate (pointing downwards) and D_{F} is the depth of the formation below the sea floor. The value of σ_h can be used as an estimate of the lost-circulation pressure, as is customary in drilling practice (Section 5.7).

Under the assumption of normal pore pressure regime, the pore pressure at depth D_{F} below the sea floor is given by

$$P_p = \rho_w g (D_{\text{SW}} + D_{\text{F}}) \qquad [5.33]$$

The operational drilling margin is then given by

$$\sigma_h - P_p = \left(\frac{d\sigma_h}{dz} - \rho_w g\right) D_{\text{F}} \qquad [5.34]$$

Since the riser is filled with mud, the static mud weight window is obtained by dividing the BHP window (Eq. [5.34]) by the sum of the sea water depth D_{SW}, the formation depth D_{F} and the air gap between the sea surface and the rig floor D_{AG} (The reader is encouraged to consult Ref. [23] for a review of reference depths.):

$$\Delta\gamma = \frac{\sigma_h - P_p}{D_{\text{F}} + D_{\text{SW}} + D_{\text{AG}}} = \left(\frac{d\sigma_h}{dz} - \rho_w g\right) \frac{1}{1 + (D_{\text{SW}} + D_{\text{AG}})/D_{\text{F}}} \qquad [5.35]$$

Eq. [5.35] shows that for a given formation depth, D_{F}, the static mud weight window is narrower at locations with greater sea water depth, D_{SW}. If the window becomes prohibitively narrow, additional casing points may be required in order to reach the target depth.

In chapter "Preventing Lost Circulation," we will see how MPD can be used to prevent lost circulation in deepwater drilling.

5.11 LOST CIRCULATION IN HIGH-PRESSURE HIGH-TEMPERATURE WELLS

Lost circulation is exacerbated in high-pressure high-temperature (HPHT) wells due to the following reasons:

- reduced drilling margins;
- deterioration of mud properties at elevated temperatures and pressures.

Reduced drilling margin is a direct result of high pore pressure. In an overpressured formation, the lower operational bound of the BHP is abnormally high. The BHP must stay above the formation pore pressure in order to avoid a kick. This narrows down the window between the minimum and the maximum operational BHP bounds. HPHT formations suffering from reduced drilling margins are found, for example, in the Gulf of Mexico and in the North Sea.

Lost-circulation issues in HPHT wells can be aggravated by poor thermal stability of the drilling fluid. Elevated temperatures make the drilling fluid thinner, reducing its ability to keep solids in suspension. The heavier solids, such as barite, settle in the annulus. This creates variation in the mud weight along the well. An example of barite sagging in an HPHT well drilled in the Gulf of Mexico was described in Ref. [33]. Solids accumulated in the lower part of the inclined section and caused a static mud weight of 19.5 lbm/gal, while the fluid in the upper part of the section, depleted of barite, had the static mud weight of 12.1 lbm/gal.

Lost circulation in HPHT wells may be aggravated by cooling-induced reduction of the hoop stress around the well. This effect, described in chapter "Stresses in Rocks," is a consequence of the relatively cold mud coming into contact with hotter rocks. The magnitude of the hoop stress reduction is given by Eq. [2.44]. Given the already narrow drilling margin in an HPHT well, the cooling effect will reduce it even further.

5.12 LOST CIRCULATION IN DEVIATED AND HORIZONTAL WELLS

Lost circulation is exacerbated in deviated and horizontal wells due to the following reasons:

- increased stress anisotropy;
- increased frictional pressure loss in the annulus as the well becomes longer;
- poor hole cleaning.

Consider an inclined wellbore drilled in a formation with extensional stress regime and isotropic horizontal stresses: $\sigma_v > \sigma_H = \sigma_h$. The in situ normal stresses acting in the plane perpendicular to the wellbore axis are given by $\sigma_H^{(n)} = \sigma_H \cos^2 i + \sigma_v \sin^2 i$ and $\sigma_h^{(n)} = \sigma_H$, where i is the angle between the wellbore axis and the vertical direction. For a vertical well ($i = 0°$), $\sigma_H^{(n)} = \sigma_h^{(n)} = \sigma_H$. For a deviated well, the value of $\sigma_H^{(n)}$ is higher than σ_H because $\sigma_v > \sigma_H$. The value of $\sigma_h^{(n)}$ is the same for a deviated well and a vertical well, $\sigma_h^{(n)} = \sigma_H$. Let us assume, as an example, that perfect filter cake is deposited on the borehole wall, and the wall is thus impermeable. The FIP is then given by $\left(3\sigma_h^{(n)} - \sigma_H^{(n)} - P_p + T_0 \right)$. Here, P_p is the pore pressure in the rock behind the filter cake; T_0 is the tensile strength of the rock. Since $\sigma_H^{(n)}$ increases with well inclination, while $\sigma_h^{(n)}$ remains the same, the FIP value decreases with inclination. A horizontal well will have the lowest FIP.

Thus, it is easier to generate an induced fracture while drilling a deviated well than when drilling a vertical well. The FPP might, however, be approximately the same in an inclined well and in the vertical well since it is largely determined by the minimum far-field in situ stress. The minimum far-field in situ stress is unaffected by the well inclination.

Increasing stress anisotropy makes the wellbore more unstable. The lower bound of the drilling window therefore increases with well inclination. The drilling window in a deviated well might be reduced to the point where it closes completely (Fig. 5.17).

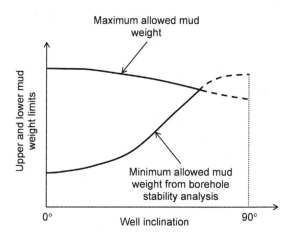

FIGURE 5.17 Effect of well inclination on the mud weight window in a difficult deviated well. The upper bound is chosen as FIP or FBP here and is thus affected by the well inclination. The lower bound is determined by the borehole stability. *Schematic plot based on Kristiansen TG, Mandziuch K, Heavey P, Kol H. Minimizing drilling risk in extended-reach wells at Valhall using geomechanics, geoscience and 3D visualization technology. SPE/IADC paper 52863 presented at the 1999 SPE/IADC Drilling Conference held in Amsterdam, Holland, 9–11 March 1999.*

In addition to the reduction of the drilling margin described above, there is another reason for lost circulation being more severe in deviated and horizontal wells. Consider a horizontal well as an example. In the horizontal section, the length of the well increases while the true vertical depth (TVD) remains approximately the same. The constant TVD means that there might be as good as no change in the in situ stresses along the horizontal section. Of course, in reality, in situ stresses will vary somewhat across the reservoir because of geological conditions or depletion. There will however be no regular increase of in situ stresses that would normally be expected in a vertical well. Constant in situ stresses imply that the fracturing pressure remains approximately constant along the horizontal section, or fluctuates only due to geological heterogeneity of the formation. At the same time, the BHP in the well increases with the length of the horizontal section since the APL increases. This means that, at some point, the BHP will exceed the fracturing pressure, inducing mud losses.

The two effects—ie, the increased stress anisotropy and increased APL—both contribute to increased losses in deviated and horizontal wells. The increased stress anisotropy facilitates fracture initiation and formation breakdown by reducing FIP and FBP. As we have seen earlier in this chapter, reduced FIP and FBP per se do not automatically imply severe mud losses since the induced fracture might still remain in the near-well area. However, increased BHP (due to APL) will force the fracture to propagate deeper into the formation than it otherwise would. This will increase mud losses.

So far in this Section, we have considered only situations with isotropic horizontal in situ stresses, ie, $\sigma_H = \sigma_h$. What if the in situ stress state is fully anisotropic; eg, $\sigma_v > \sigma_H > \sigma_h$? In this case, the FIP value for a horizontal wellbore depends on the wellbore orientation with regard to σ_H, σ_h. Assume, again, that perfect filter cake is deposited on the wellbore wall. If the wellbore axis is parallel to the maximum horizontal stress, σ_H, the FIP is given by $(3\sigma_h - \sigma_v - P_p + T_0)$. If the wellbore axis is parallel to the minimum horizontal stress, σ_h, the FIP is given by $(3\sigma_H - \sigma_v - P_p + T_0)$. The FIP thus depends on the wellbore orientation in the horizontal plane, and is smaller for a wellbore parallel to σ_H in this case. Even though smaller FIP does not necessarily imply larger mud losses, it increases the risk of fracturing and makes the overall situation worse if the BHP is sufficient to create long fractures after the breakdown. This illustrates the role of stress anisotropy and wellbore orientation in lost circulation.

In addition to the two first reasons for severe lost-circulation problems in horizontal and deviated wells (ie, increased stress anisotropy and increased APL), there is another one: poor hole cleaning. Eccentric positioning of the drillpipe in a horizontal or deviated well impairs cuttings transport in the narrow part of the annulus, resulting in a bed buildup. Cuttings production and accumulation can be enhanced by borehole instabilities caused by the BHP being below the lower operational bound. In addition to increasing the amount of cuttings, borehole instabilities increase the borehole cross-section, thereby reducing the annular fluid velocity. All these processes—ie, increased amount of cuttings, reduced annular velocity, and eccentric pipe—contribute to poor cleaning of the hole. Poor cleaning may eventually result in a packoff of the drillstring. A packoff may then cause spikes in the bottomhole circulating pressure, with induced fracturing and mud losses as a consequence [33].

Deviated and horizontal (extended reach) wells are often also HPHT wells. Lost circulation can therefore be additionally aggravated by the factors common to HPHT wells (Section 5.11). Some of those factors can even be enhanced in deviated wells. For instance, barite sagging is facilitated by a mechanism described in Ref. [33]: insufficient gel strength caused by elevated temperature promotes barite sagging. Barite settles onto the lower side of the annulus in the deviated section and slides down the well. The fluid depleted of barite is accumulated in the upper part of the well cross-section. Since this fluid is relatively light, it flows up the well. This flow increases the shear rate and reduces the mud viscosity because mud's apparent viscosity decreases with shear rate. The result is a thinner fluid which further facilitates sagging.

In deepwater drilling, lost circulation issues in horizontal and deviated wells may become exacerbated by varyied water depth. For instance, a variation in the sea water depth from 4420 ft (1347 m) to 6500 ft (1981 m) was reported over a lateral distance of 2 miles for an extended reach well drilled in the Gulf of Mexico [37]. The differences in the sea water depth affect the maximum allowed mud weight by reducing the fracture gradient in the well sections drilled under deeper sea (Eq. [5.35]). If this effect becomes substantial, it may become impossible to drill the entire interval with the same mud weight without lost circulation in some parts of the hole, or wellbore instability, or fluid influx in other parts. Avoiding drilling troubles will require additional casing points, which impacts both the well cost and the well diameter at the target depth.

5.13 LOST CIRCULATION IN GEOTHERMAL DRILLING

Lost circulation is exacerbated in geothermal drilling due to the following reasons:

- geothermal reservoirs are often underpressured;
- geothernal reservoirs typically have natural fractures of large aperture (on the order of centimeter).

Low pore pressure in geothermal reservoirs means that there is significant overbalance unless, for example, aerated muds or drilling with air are employed.

Natural fractures are the primary pathways for the reservoir fluids to the well. Therefore large fracture aperture is one of the factors making a geothermal reservoir commercially viable. For this reason, the well trajectories are normally chosen so as to intersect as many natural fractures as possible. However, large apertures and low fluid pressure in the fractures are a recipe for lost circulation. Apertures in excess of 5 mm, not uncommon in geothermal fields, represent a serious challenge for currently available LCMs.

5.14 EFFECT OF BASE FLUID ON LOST CIRCULATION

The type of the base fluid affects lost circulation in permeable rocks; eg, sandstones. In particular, laboratoty experiments show the following effects [22,40,41]:

- The type of the base fluid has virtually no effect on the FIP in wells without preexisting cracks (notches, flaws).
- The type of the drilling fluid has an effect on the FPP. The FPP is higher with WBM than with OBM.
- The FRP is higher with WBM than with OBM.
- During fracture propagation, pressure fluctuations with WBM are larger than with OBM. The pressure versus time curve during fracture propagation has a characteristic "sawtooth shape" with WBM.
- After breakdown, the fracture propagation with OBM is unstable (the wellbore pressure drops abruptly) while a stable fracture growth is obtained with WBM.
- Induced fractures may "heal" in the presence of WBM.

As a result of the above, lost-circulation problems are more common when drilling with OBM and SBM than with WBM [27]. Permeable formations that could be drilled with WBM with no or partial losses may often give rise to severe losses when drilled with SBM or OBM.

In low-permeability rocks (eg, shale), there are no clear or systematic differences between OBM and WBM with respect to the values of FIP, FPP, or FRP.

The effect of mud type and rock permeability on FIP, FPP, and FRP is summarized in Table 5.3. Subscripts in Table 5.3 denote the type of mud. For instance, FIP_{OBM} refers to the FIP obtained with an OBM.

It was emphasized in Refs. [21−23] that the differences between WBM and OBM are due to their different filtration properties. Therefore, WBM in Table 5.3 represents a high-fluid-loss mud rather than a particular mud composition. Similarly, OBM signifies a low-fluid-loss mud. Spurt loss and filtration rate of WBM are typically higher, which affects the filter cake buildup inside the fracture in a permeable rock. In such a rock, improved ability of WBM to build the filter cake and seal the fracture increases the values of FBP, FPP, and FRP. Better filtration capacity of WBM as opposed to OBM entails that the fracture tip is better protected from the wellbore pressure. As a result, the formation breakdown requires higher wellbore pressure, which

TABLE 5.3 Effect of Mud Type on Lost Circulation

Formation Permeability	Fracture Initiation Pressure (Intact Wellbore Wall)	Fracture Propagation Pressure	Fracture Reopening Pressure
High-permeability rock (eg, sandstone)	$FIP_{OBM} \approx FIP_{WBM}$	$FPP_{OBM} < FPP_{WBM}$; sawtooth shape of pressure vs. time curve with WBM	$FRP_{OBM} < FRP_{WBM}$; $FRP_{OBM} \approx FPP_{OBM}$
Low-permeability rock (eg, shale)	$FIP_{OBM} \approx FIP_{WBM}$	$FPP_{OBM} \approx FPP_{WBM}$; no sawtooth shape of pressure vs. time curve with WBM	$FRP_{OBM} \approx FRP_{WBM}$; $FRP_{OBM} \approx FPP_{OBM}$

Based on Onyia EC. Experimental data analysis of lost-circulation problems during drilling with oil-based muds. SPE Drill Completion 1994;9(1):25–31.

is perceived as an increase of FBP with WBM. The effect of WBM on FPP and FRP has a similar explanation.

As evident from Table 5.3, the difference between high-fluid-loss and low-fluid-loss muds is most pronounced in high-permeability rocks. Higher values of FPP and FRP obtained with WBMs support the use of water-based lost circulation pills when losses are experienced in high-permeability rocks drilled with OBM [41].

Water-base drilling fluids (or, in general, muds with better filtration properties) thus can reduce the lost circulation risk, all the rest being equal. Consequently, the upper bound of the operational mud weight window could be set higher with WBM than with nonaqueous fluids [40].

Another difference between WBM and OBM is the behavior of the dehydrated zone as the fracture propagates. With WBMs, the size of the dehydrated zone increases as the fracture propagates. With OBMs, the size of the dehydrated zone remains approximately constant during fracture propagation. The size of the noninvaded zone depends on the mud type, too, and is greater with WBM than with OBM [22].

5.15 IDENTIFYING THIEF ZONES THROUGH MEASUREMENT WHILE DRILLING AND LOGGING

Knowing the location and properties of the thief zone is required for combatting lost circulation. Logging and other measurements (both surface and downhole) can be used to identify the type of losses, to locate the thief zones (eg, fractures), and to evaluate some of their properties. Examples of such measurements, discussed in this section, are the equivalent circulating density (ECD) measurements performed as part of measurement while drilling (MWD) and the propagation resistivity log.

Measuring the ECD while drilling may provide an indication of drilling-induced fracturing. At the onset of mud circulation, the ECD increases more rapidly if the formation is intact, and exhibits a slower, more gradual increase if the formation is fractured (Fig. 5.18). The gradual rather than steep increase of the ECD in the latter case is caused by the fluid flowing into the fractures. Conversely, when circulation is stopped, the ECD drops rapidly if the formation is intact, but may decrease gradually if the formation is fractured (Fig. 5.19). The gradual rather than steep decrease of the ECD in the latter case is caused by some of the lost fluid flowing from the fractures back into the well. The two characteristic shapes of the ECD versus time curve have been called "square response" (intact formation, Figs. 5.18A and 5.19A) and "exponential tails" (fractured formation, Figs. 5.18B and 5.19B) [42,43].

A propagation resistivity log can be helpful at finding fractures and quantifying their dimensions. This method is based on two measurements: one providing information about the near-well area in the immediate vicinity of the well; the other (low-frequency signal) penetrating deeper into the formation [43]. Electric current induced by the logging tool flows in the circumferential direction around the well. Thus, if a radial fracture is present within the depth of investigation and is filled with a high-resistivity SBM, it will boost the measured resistivity. Separation of the curves measured at the two depths

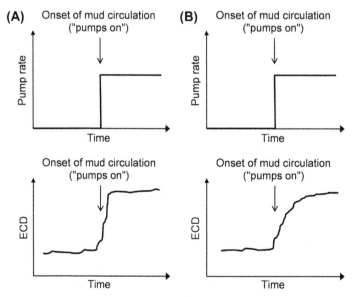

FIGURE 5.18 Pump rate and equivalent circulating density (ECD) versus time when mud circulation is resumed: (A) the well is in an intact formation; (B) the near-well area is fractured.

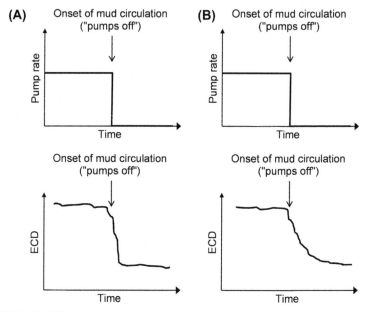

FIGURE 5.19 ECD versus time when mud circulation is stopped: (A) the well is in an intact formation; (B) the near-well area is fractured.

indicates a short fracture (or fractures) in the immediate vicinity of the well and can therefore be used to locate the lost-circulation zone [42]. The length of the interval where the two curves are separated provides an estimate of the fracture size along the well [43] (Fig. 5.20). Increase in resistivity in both curves, without separation, indicates a long fracture penetrating deep into the formation in the radial direction from the well (Fig. 5.21). The fracture dimension along the radius equals in this case at least the depth of investigation of the deeper resistivity measurement.

Additional insight into fractures in shale can be obtained by time-lapse resistivity measurements performed at different ECD values. Increasing ECD leads to fracture reopening, which is then observed as an increase in the resistivity along the thief zone (Fig. 5.22). Conversely, decreasing ECD leads to fracture closing, with a corresponding decrease in the resistivity (Fig. 5.23) [43].

Temperature survey is another method available for locating losses. Temperature profile along the open hole is logged several hours after the circulation was stopped. The circulation is then resumed, and the temperature profile is measured again. The changes (temperature discontinuities) between the two profiles indicate where the mud goes during circulation [44]. Temperature logging is used to locate thief zones in geothermal wells [45].

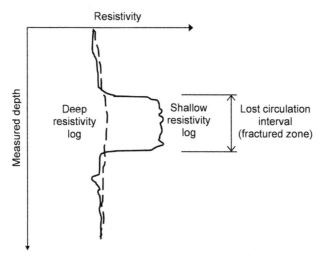

FIGURE 5.20   Deep and shallow resistivity logs in the presence of a short fracture. Wellbore is drilled in shale with synthetic-base mud. Separation of the two curves indicates the fracture or a fractured zone filled with high-resistivity mud. *Solid line*: shallow resistivity log. *Dashed line*: deep resistivity log. *Schematic plot based on Bratton TR, Rezmer-Cooper IM, Desroches J, Gille Y-E, Li Q, McFayden M. How to diagnose drilling induced fractures in wells drilled with oil-based muds with real-time resistivity and pressure measurements. SPE/IADC paper 67742 presented at the SPE/ IADC Drilling Conference held in Amsterdam, The Netherlands, 27 February–1 March 2001.*

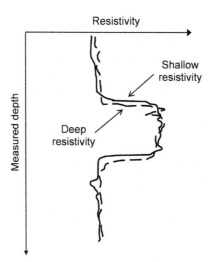

FIGURE 5.21   Deep and shallow resistivity logs in the presence of a deep fracture. Wellbore is drilled in shale with synthetic-base mud. Separation of the two curves indicates the fracture or a fractured zone filled with high-resistivity mud. *Solid line*: shallow resistivity log. *Dashed line*: deep resistivity log. *Schematic plot based on Bratton TR, Rezmer-Cooper IM, Desroches J, Gille Y-E, Li Q, McFayden M. How to diagnose drilling induced fractures in wells drilled with oil-based muds with real-time resistivity and pressure measurements. SPE/IADC paper 67742 presented at the SPE/IADC Drilling Conference held in Amsterdam, The Netherlands, 27 February–1 March 2001.*

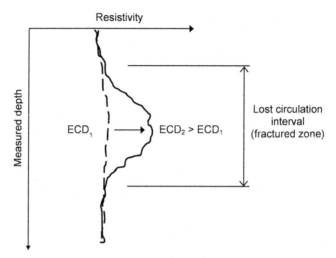

FIGURE 5.22   Change in the resistivity log with increasing ECD. Wellbore is drilled in shale with a synthetic-base mud. Fracture is not visible in the resistivity log at low ECD (*dashed line*, ECD_1). The fracture opens up and becomes evident in the resistivity log when the ECD is increased to ECD_2 (*solid line*). *Schematic plot based on Bratton TR, Rezmer-Cooper IM, Desroches J, Gille Y-E, Li Q, McFayden M. How to diagnose drilling induced fractures in wells drilled with oil-based muds with real-time resistivity and pressure measurements. SPE/IADC paper 67742 presented at the SPE/IADC Drilling Conference held in Amsterdam, The Netherlands, 27 February—1 March 2001.*

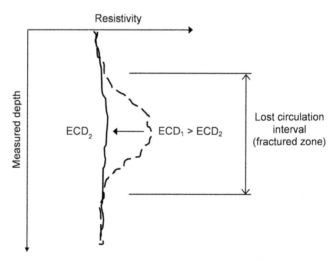

FIGURE 5.23   Change in the resistivity log with decreasing ECD. Wellbore is drilled in shale with a synthetic-base mud. Fracture is evident in the resistivity log at high ECD (*dashed line*, ECD_1). The fracture closes and disappears from the resistivity log when the ECD is decreased to ECD_2 (*solid line*). *Schematic plot based on Bratton TR, Rezmer-Cooper IM, Desroches J, Gille Y-E, Li Q, McFayden M. How to diagnose drilling induced fractures in wells drilled with oil-based muds with real-time resistivity and pressure measurements. SPE/IADC paper 67742 presented at the SPE/IADC Drilling Conference held in Amsterdam, The Netherlands, 27 February—1 March 2001.*

In addition to logging, indirect evidence available at the rig can be used to locate the loss zone. For instance, losses are believed to originate at the drill bit in the following situations [46]:

- if they are observed when drilling ahead;
- if they are accompanied with a significant change in the rate of penetration, torque, or vibration;
- if the loss occurs while entering a fractured, vugular, or high-permeability zone known from geological data.

Losses are believed to occur not necessarily at the drill bit under the following circumstances [46]:

- losses are observed while tripping or increasing the mud weight;
- losses are observed in a shut-in or killed well.

The greatest challenge of fracture characterization is the evaluation of the fracture aperture. At present, there are no techniques that could be routinely used evaluate this parameter accurately.

5.16 INTERPRETATION AND DIFFERENTIAL DIAGNOSIS OF LOSSES

Identifying the type of losses and the location of the loss zone is necessary; eg, for placing an LCM pill [47]. If high-frequency measurements of flow rates in and out of the well are available, they can be used as an indicator of whether losses are due to high matrix permeability, a vugular zone, a natural fracture, or an induced fracture. In the absence of such measurements, indirect indicators can be used. For instance, losses caused by induced fractures are believed to be much more sensitive to changes in the wellbore pressure than losses caused by natural fractures. This is so because, in the case of a natural fracture, losses often occur into a mechanically closed fracture (fracture faces are in contact with each other at contact spots created by asperities), as long as the wellbore pressure is below the FRP. An induced fracture, on the contrary, is opened both hydraulically and mechanically (faces are not in contact with each other). Increasing the net pressure may therefore increase their aperture and permeability significantly (chapter: Natural Fractures in Rocks).

Losses into natural fractures start suddenly and are not usually accompanied with changes in drilling parameters, such as the stand pipe pressure. Occasionally, an increase in the rate of penetration may be observed.

Losses into natural fractures do not necessarily require a pressure surge. All that is required for such losses to start is the availability of a sufficiently wide fracture for mud invasion and the BHP in excess of the formation fluid pressure (ie, overbalance).

Induced fractures are a consequence of the downhole pressure becoming higher than the fracturing pressure. Such a BHP increase can result, for example, from a pressure surge when the pumps are turned on. Other in-dicators of losses caused by induced fractures are variations in the stand pipe pressure, torque, and weight on bit [5]. As mentioned above, the aperture of an induced fracture is quite sensitive to variations in the wellbore pressure, and a small increase in the overbalance may induce large losses.

Losses due to induced fractures can be caused by poor hole cleaning. Poor hole cleaning may build up mud rings on the drill string. As a result, the ECD may increase to the level that exceeds the fracturing pressure. A fracture will then be induced in the formation [5].

Another hint at losses being caused by induced fractures can be obtained by comparing the BHP to the expected fracturing pressure in the formation. If the wellbore pressure stays below the fracturing pressure, and at the same time the loss volume is not extremely sensitive to the wellbore pressure changes, the losses are most likely caused by natural fractures [6].

Losses into rubble zones can be identified by cavings analysis or resistivity data. As pointed out in Ref. [37], timely and accurate identification of rubble zones is crucial for combatting lost circulation and wellbore instabilities. Similar to rubble zones, formations around faults are prone to lost circulation. Identifying such zones is possible from geological and seismic data.

Effective application of particulate LCMs in naturally fractured formations requires that the aperture of the fractures is taken into account when choosing the size of LCM particles. The particle size optimal for bridging a natural fracture is around 40% of the fracture aperture. Other numbers have been quoted as well; eg, 15−33% [8]. Designing the LCM based on these criteria requires that information (or an educated guess) about the fracture aperture is available. Such information is difficult to obtain. In particular, the fracture aperture cannot normally be evaluated from image logs because of insufficient resolution. Readings from image logs tend to overestimate the fracture aper-ture [31].

It has been proposed to use mud loss data to evaluate the aperture of natural fractures [4,6]. According to this technique, high-frequency measurements of the flow rate in and out of the well are processed and interpreted using the theory reviewed in Section 5.4. Matching the "flow rate versus time" curve measured at the rig with the type curves similar to those shown in Fig. 5.4 could yield an estimate of the fracture aperture. Furthermore, the fracture aperture can be estimated from the total volume of the lost fluid according to Eqs. [5.5]−[5.7] [48].

This technique has been used by some operators to characterize fractures [4−6,8,48]. The ability to directly detect *hydraulically opened* (ie, *conductive*) fractures is a significant advantage of this technique. This ability distinguishes it from other fracture evaluation techniques (cores, image logs). Widespread application of mud-loss-based formation evaluation techniques requires, however, that several practical and theoretical challenges are addressed first. In particular:

- The acquisition of high-frequency flow rate data requires installation of flowmeters which, in turn, entails modifications to the mud hydraulic system.
- The interpretation of the collected high-frequency data has so far been based on simplistic models developed in Refs. [8,9]. The assumptions made in these models may not always be true. This will produce errors in the predicted fracture apertures.

Interpretation of mud losses and their mechanisms is improved by combining several methods, including core analysis, logging, monitoring of operational drilling parameters, pit levels, and flow rate data.

5.17 SUMMARY

Combatting lost circulation requires that the mechanisms of losses are understood and can be identified during drilling. Losses are usually attributed to one of the four mechanisms (or a combination thereof): high-permeability matrix, vugular formations, natural fractures, and drilling-induced fractures. The variety of the mechanisms means that lost-circulation pressure cannot be uniquely defined (Fig. 5.24). In some cases, severe losses may occur when the BHP exceeds the formation pore pressure. In other cases, BHPs in excess of the minimum in situ stress can be sustained. The upper bound of the BHP is often set equal to the minimum in situ stress. This strategy prevents massive losses if lost circulation is due to induced or narrow natural fractures (Fig. 5.24).

The occurrence and the amount of losses may depend, apart from in situ stresses, on the drilling fluid rheology, elastic properties of the rock, the presence, orientation, and aperture of natural fractures, etc. Since most of these factors cannot be quantified before the well is actually drilled, setting the upper operational bound of BHP equal to the minimum in situ stress is a reasonable strategy. However, losses might occur also when the wellbore pressure is below the minimum in situ stress, or they might not occur even when the wellbore pressure is above the minimum in situ stress. This conundrum is recognized in the drilling and rock mechanics communities. For instance, Buechler et al. emphasized that in naturally fractured reservoirs or in formations with multiple shale—sand interfaces the lost-circulation pressure may be equal to the FRP, FPP, leakoff pressure, or formation breakdown pressure, depending on the presence (or absence) and orientation of natural fractures and rock interfaces [19].

Losses are exacerbated in certain types of formations, in particular in depleted reservoirs, deepwater wells, HPHT wells, deviated, and horizontal

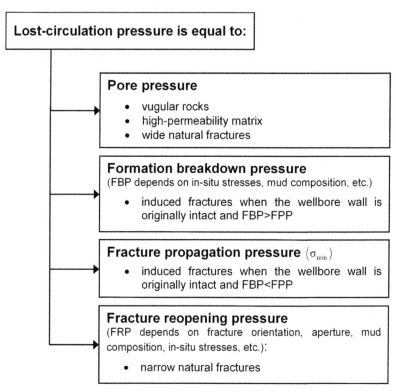

FIGURE 5.24 Lost-circulation pressure ambiguity.

TABLE 5.4 Mechanisms and Diagnostics of Lost Circulation

Mechanism of Lost Circulation	Diagnostic Features
High-porosity rock	• Losses start gradually • Loss flow rate increases gradually and may then gradually decrease as filter cake builds up
Vugular formation	• Losses start suddenly • Severe or total losses • Impossible to cure with LCM • Losses in specific types of formations; eg, carbonates (karst) • Drill bit may drop a few meters when it hits the vug
Natural fractures	• Losses start suddenly as fractures are intersected by the wellbore
Drilling-induced fractures	• Losses often accompany pressure surges (eg, when running pipe in hole or starting the pump)

wells. Geothermal drilling is known for severe losses caused by large fracture apertures and underpressured formations. Being aware of these problems in advance may help one implement preventive measures in such wells.

Implementation of preventive measures requires that losses are properly diagnosed, both in terms of their mechanism and location of the thief zone. Ideally, properties of fractures and vugs causing losses should be quantified as well; eg, the aperture of natural or induced fractures.

Table 5.4 provides some examples of diagnostic features of lost circulation mechanisms. In the next chapter, we will see how preventive measures help to avoid or mitigate losses caused by different mechanisms.

REFERENCES

[1]   Nelson EB, Guillot D. Well cementing. 2nd ed. Schlumberger; 2006. 773 p.

[2]   Alsaba M, Nygaard R, Saasen A, Nes O-M. Lost circulation materials capability of sealing wide fractures. SPE paper 170285 presented at the SPE Deepwater Drilling and Completions Conference held in Galveston, Texas, USA, 10—11 September 2014.

[3]   DrillingSpecialtiesCompany. Lost circulation guide. 2014.

[4]   Sanfillippo F, Brignoli M, Santarelli FJ, Bezzola C. Characterization of conductive fractures while drilling. SPE paper 38177 presented at the 1997 SPE European Formation Damage Conference held in The Hague, The Netherlands, 2—3 June 1997.

[5]   Beda G, Carugo C. Use of mud microloss analysis while drilling to improve the formation evaluation in fractured reservoir. SPE paper 71737 presented at the 2001 SPE Annual Technical Conference and Exhibition held in New Orleans, Louisiana, 30 September—3 October 2001.

[6]   Dyke CG, Wu B, Milton-Tayler D. Advances in characterizing natural-fracture permeability from mud-log data. SPE Form Eval 1995;10(3):160—6.

[7]   Kumar A, Savari S, Jamison DE, Whitfill DL. Application of fiber laden pill for controlling lost circulation in natural fractures. AADE-11-NTCE-19 paper presented at the 2011 AADE National Technical Conference and Exhibition held at Hilton Houston North Hotel, Houston, Texas, April 12—14, 2011.

[8]   Liétard O, Unwin T, Guillot DJ, Hodder MH. Fracture width logging while drilling and drilling mud/loss-circulation-material selection guidelines in naturally fractured reservoirs. SPE Drill Completion 1999;14(3):168—77.

[9]   Majidi R, Miska SZ, Yu M, Thompson LG, Zhang J. Quantitative analysis of mud losses in naturally fractured reservoirs: the effect of rheology. SPE Drill Completion 2010;25(4):509—17.

[10] Majidi R, Miska SZ, Ahmed R, Yu M, Thompson LG. Radial flow of yield-power-law fluids: numerical analysis, experimental study and the application for drilling fluid losses in fractured formations. J Pet Sci Eng 2010;70:334—43.

[11] Ghalambor A, Salehi S, Shahri MP, Karimi M. Integrated workflow for lost circulation prediction. SPE paper 168123 presented at the SPE International Symposium and Exhibition on Formation Damage Control held in Lafayette, Louisiana, USA, 26—28 February 2014.

[12] Civan F, Rasmussen ML. Further discussion of fracture width logging while drilling and drilling mud/loss-circulation-material selection guidelines in naturally fractured reservoirs. SPE Drill Completion December 2002:249—50.

[13] Lavrov A, Tronvoll J. Modeling mud loss in fractured formations. SPE paper 88700 presented at the Abu Dhabi International Conference and Exhibition held in Abu Dhabi, UAE, 10−13 October 2004.

[14] Lavrov A. Newtonian fluid flow from an arbitrarily-oriented fracture into a single sink. Acta Mech 2006;186:55−74.

[15] Zeilinger S, Dupriest F, Turton R, Butler H, Wang H. Utilizing an engineered particle drilling fluid to overcome coal drilling challenges. IADC/SPE paper 128712 presented at the 2010 IADC/SPE Drilling Conference and Exhibition held in New Orleans, Louisiana, USA, 2−4 February 2010.

[16] Iwashita D, Morita N, Tominaga M. Shear-type borehole wall shifts induced during lost circulations. SPE Drill Completion 2008;23(3):301−13.

[17] Dupriest FE. Fracture closure stress (FCS) and lost returns practices. SPE/IADC paper 92192 presented at the SPE/IADC Drilling Conference held in Amsterdam, The Netherlands, 23−25 February 2005.

[18] Sanders MW, Young S, Friedheim JE. Development and testing of novel additives for improved wellbore stability and reduced losses. AADE-08-DF-HO-19 paper presented at the 2008 AADE Fluids Conference and Exhibition held at the Wyndam Greenspoint Hotel, Houston, Texas, April 8−9, 2008.

[19] Buechler S, Ning J, Bharadwaj N, Kulkarni K, DeValve C, Elsborg C, et al. Root cause analysis of drilling lost returns in injectite reservoirs. SPE paper 174848 presented at the SPE Annual Technical Conference and Exhibition held in Houston, Texas, USA, 28−30 September 2015.

[20] Ito T, Evans K, Kawai K, Hayashi K. Hydraulic fracture reopening pressure and the estimation of maximum horizontal stress. Int J Rock Mech Min Sci 1999;36:811−26.

[21] Morita N, Fuh G-F, Black AD. Borehole breakdown pressure with drilling fluids − II. Semi-analytical solution to predict borehole breakdown pressure. Int J Rock Mech Min Sci Geomech Abstr 1996;33(1):53−69.

[22] Morita N, Black AD, Fuh G-F. Borehole breakdown pressure with drilling fluids − I. Empirical results. Int J Rock Mech Min Sci Geomech Abstr 1996;33(1):39−51.

[23] Fjær E, Holt RM, Horsrud P, Raaen AM, Risnes R. Petroleum related rock mechanics. 2nd ed. Amsterdam: Elsevier; 2008. 491 p.

[24] Guo Q, Cook J, Way P, Ji L, Friedheim JE. A comprehensive experimental study on wellbore strengthening. IADC/SPE paper 167957 presented at the 2014 IADC/SPE Drilling Conference and Exhibition held in Fort Worth, Texas, USA, 4−6 March 2014.

[25] Økland D, Gabrielsen GK, Gjerde J, Sinke K, Williams EL. The importance of extended leak-off test data for combatting lost circulation. SPE/ISRM paper 78219 presented at the SPE/ISRM Rock Mechanics Conference held in Irving, Texas, 20−23 October 2002.

[26] Lavrov A, Larsen I, Holt RM, Bauer A, Pradhan S. Hybrid FEM/DEM simulation of hydraulic fracturing in naturally-fractured reservoirs. ARMA paper 14−7107 presented at the 48th US Rock Mechanics/Geomechanics Symposium held in Minneapolis, MN, USA, 1−4 June 2014.

[27] Tare UA, Whitfill DL, Mody FK. Drilling fluid losses and gains: case histories and practical solutions. SPE paper 71368 presented at the 2001 SPE Annual Technical Conference and Exhibition held in New Orleans, Louisiana, 30 September−3 October 2001.

[28] Adachi J, Bailey L, Houwen OH, Meeten GH, Way PW, Schlemmer RP. Depleted zone drilling: reducing mud losses into fractures. IADC/SPE paper 87224 presented at the IADC/SPE Drilling Conference held in Dallas, Texas, U.S.A., 2−4 March 2004.

[29] Bourgoyne Jr AT, Millheim KK, Chenevert ME, Young Jr FS. Applied drilling engineering. Richardson: Society of Petroleum Engineers; 1991.

[30] Ram Babu D. Effect of P$-\rho-$T behavior of muds on loss/gain during high-temperature deep-well drilling. J Pet Sci Eng 1998;20(1$-$2):49$-$62.

[31] Ameen MS. Fracture and in-situ stress patterns and impact on performance in the Khuff structural prospects, eastern offshore Saudi Arabia. Mar Pet Geol 2014;50:166$-$84.

[32] Ozdemirtas M, Babadagli T, Kuru E. Experimental and numerical investigations of borehole ballooning in rough fractures. SPE Drill Completion 2009;24(2):256$-$65.

[33] Gradishar J, Ugueto G, van Oort E. Setting free the bear: the challenges and lessons of the Ursa A-10 deepwater extended-reach well. SPE Drill Completion 2014;29(2):182$-$93.

[34] Wang H, Sweatman R, Engelman B, Deeg W, Whitfill D, Soliman M, et al. Best practice in understanding and managing lost circulation challenges. SPE Drill Completion 2008;23(2):168$-$75.

[35] Kartevoll M. Drilling problems in depleted reservoirs [MSc thesis]. Stavanger: University of Stavanger; 2009.

[36] Hettema MHH, Bostrøm B, Lund T. Analysis of lost circulation during drilling in cooled formations. SPE paper 90442 presented at the SPE Annual Technical Conference and Exhibition held in Houston, Texas, U.S.A., 26$-$29 September 2004.

[37] Willson SM, Edwards S, Heppard PD, Li X, Coltrin G, Chester DK, et al. Wellbore stability challenges in the deep water, Gulf of Mexico: case history examples from the Pompano Field. SPE paper 84266 presented at the SPE Annual Technical Conference and Exhibition held in Denver, Colorado, U.S.A., 5$-$8 October, 2003.

[38] Rehm B, Schubert J, Haghshenas A, Paknejad AS, Hughes J, editors. Managed pressure drilling. Houston, Texas: Gulf Publishing Company; 2008.

[39] Kristiansen TG, Mandziuch K, Heavey P, Kol H. Minimizing drilling risk in extended-reach wells at Valhall using geomechanics, geoscience and 3D visualization technology. SPE/IADC paper 52863 presented at the 1999 SPE/IADC Drilling Conference held in Amsterdam, Holland, 9$-$11 March 1999.

[40] Growcock FB, Patel AD. The revolution in non-aqueous drilling fluids. Paper AADE-11-NTCE-33 presented at the 2011 AADE National Technical Conference and Exhibition held at the Hilton Houston North Hotel, Houston, Texas, April 12$-$14, 2011.

[41] Onyia EC. Experimental data analysis of lost-circulation problems during drilling with oil-based muds. SPE Drill Completion 1994;9(1):25$-$31.

[42] Caughron DE, Renfrow DK, Bruton JR, Ivan CD, Broussard PN, Bratton TR, et al. Unique crosslinking pill in tandem with fracture prediction model cures circulation losses in deepwater Gulf of Mexico. IADC/SPE paper 74518 presented at the IADC/SPE Drilling Conference held in Dallas, Texas, 26$-$28 February 2002.

[43] Bratton TR, Rezmer-Cooper IM, Desroches J, Gille Y-E, Li Q, McFayden M. How to diagnose drilling induced fractures in wells drilled with oil-based muds with real-time resistivity and pressure measurements. SPE/IADC paper 67742 presented at the SPE/IADC Drilling Conference held in Amsterdam, The Netherlands, 27 February$-$1 March 2001.

[44] Messenger JU. Lost circulation. PenWell Books; 1981. 112 p.

[45] Finger J, Blankenship D. Handbook of best practices for geothermal drilling. Sandia National Laboratories; 2010. Contract No.: SAND2010$-$6048.

[46] Power D, Ivan CD, Brooks SW. The top 10 lost circulation concerns in deepwater drilling. SPE paper 81133 presented at the SPE Latin American and Caribbean Petroleum Engineering Conference held in Port-of-Spain, Trinidad, West Indies, 27$-$30 April 2003.

[47] Droger N, Eliseeva K, Todd L, Ellis C, Salih O, Silko N, et al. Degradable fiber pill for lost circulation in fractured reservoir sections. IADC/SPE paper 168024 presented at the 2014 IADC/SPE Drilling Conference and Exhibition held in Fort Worth, Texas, USA, 4−6 March 2014.

[48] Huang J, Griffiths DV, Wong S-W. Characterizing natural-fracture permeability from mud-loss data. SPE J March 2011:111−4.

Chapter 6

Preventing Lost Circulation

Lost circulation is easier to prevent than to cure. Planning the well trajectory so as to avoid potential thief zones is the most obvious way to prevent losses. Optimization of well trajectories is facilitated by advancements in directional drilling. However, in addition to making the well more expensive, such perfect well trajectories sometimes cannot be designed at all, for instance, when accessing formations located under depleted horizons is required. A variety of loss prevention measures are available in such cases. The properties of the drilling fluid, namely its density (mud weight) and rheology, can be modified in order to reduce the static bottomhole pressure (BHP) and the annular pressure loss (APL). Wellbore strengthening can be used to increase the lost-circulation pressure of the formation, when reductions in the BHP cannot be achieved by drilling fluid optimization or when such reductions entail formation fluid influx in overpressured zones in the same interval. Casing while drilling (CWD) is another technique that is known to reduce the lost-circulation risk. Managed pressure drilling (MPD) and underbalanced drilling (UBD) may eliminate losses altogether, but they do so at the expense of extra costs. In this chapter, these and other preventive measures against lost circulation are reviewed, and their advantages and disadvantages are discussed.

6.1 IDENTIFICATION OF POTENTIAL LOSS ZONES

The most obvious way to prevent lost circulation is to avoid drilling through loss zones. This can, at least in theory, be achieved by optimizing the well trajectory. Geological analysis can help in identifying, before drilling commences, the formations prone to lost circulation (high-permeability strata, naturally fractured or cavernous rocks, etc.). For example, a study of fractures in the Khuff Formation (Saudi Arabia) revealed that fractures occur mostly in low-porosity rocks and are more common in dolomite than in limestone [1]. This type of information may help operators identify potential thief zones beforehand.

There is evidence that natural fractures sometimes occur in clusters. For instance, an analysis of image logs and cores from a granitic formation revealed that fractures were clustered in 10−20 m thick zones, with relatively consistent fracture orientation within a cluster [2]. Being aware of such fracture localization is useful when deciding on preventive measures against lost circulation.

Lost Circulation. http://dx.doi.org/10.1016/B978-0-12-803916-8.00006-6
163

Empirical evidence suggests that fracture densities are usually higher in more brittle rocks. Fractures are often associated with structural elements of high curvature; eg, folds [3]. Faults usually mean trouble as well. Faults alter in situ stresses, both in magnitude and orientation. The rock is usually weaker in and around the fault. The fracture density decreases toward the background level with distance from the fault plane. The width of the fractured zone around the fault is higher for faults with larger displacement [3]. Geological and seismic characterization of faults may therefore help in designing a well trajectory that will avoid thief zones. The well can be designed so that it either avoids the near-fault fractured zone or runs normal to the fault plane, minimizing the exposure [3]. Field cases discussed by Kristiansen et al. [3] suggest that while some faults are trouble, others are not. This unpredictability complicates the problem.

The importance of fracture prediction and characterization for successful implementation of loss prevention measures is well recognized in the industry. It is, however, acknowledged that such characterization is rarely possible [4]. One of the available tools is the microresistivity imaging tool. Microresistivity measurements characterize formation up to a certain depth beyond the borehole wall (about one inch) [5]. In overbalanced drilling, the drilling fluid penetrates to some depth into natural fractures and thereby creates electric contrast. This contrast can be sensed using the microresistivity imaging tool. These tools can be used to identify the locations and estimate the density (spacing) of natural fractures in exploration wells. The resolution of image logs (on the order of 0.1 in [5]) is, however, too low to measure accurately the aperture of conducting fractures, which may be less than a millimeter. A further limitation of microresistivity imaging is the need for conductive mud, which confines their use to the wells drilled with WBM. A microinduction imaging tool is an alternative solution that works with nonaqueous fluids [5].

Acoustic image logs can be used to identify large features, such as vugs and wide-opened fractures, on the borehole wall. Their capacity is, however, poorer than that of microresistivity image logs since acoustic image logs essentially measure only the geometry of the borehole wall [5].

Geological information and seismic data can be used to identify potential thief zones. For instance, it is commonly known that rubble zones in subsalt or around faults are prone to severe or total losses. Sand-shale interfaces often cause significant losses. As an example, Fig. 6.1 shows cumulative mud losses as a function of spatial density of such interfaces. Fig. 6.1 is based on the data reported by Buechler et al. for several North Sea wells [6].

Depleted formations are prone to losses because the pore pressure and the fracturing pressure are reduced there (chapter: Mechanisms and Diagnostics of Lost Circulation). Being aware of the locations of depleted zones helps minimize losses when drilling through such formations.

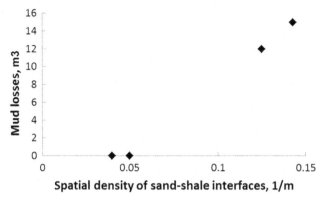

FIGURE 6.1   Mud losses caused by sand—shale interfaces as a function of spatial density of the interfaces. *Based on the data reported by Buechler S, Ning J, Bharadwaj N, Kulkarni K, DeValve C, Elsborg C, et al. Root cause analysis of drilling lost returns in injectite reservoirs. SPE paper 174848 presented at the SPE Annual Technical Conference and Exhibition held in Houston, Texas, USA; September 28—30, 2015.*

Advancements in directional drilling enable planning the well to avoid thief zones or at least minimize the exposure. Planning the well so as to avoid thief zones is, of course, not always possible. However, knowing in advance where such zones are located enables a timely deployment of other loss prevention measures; eg, wellbore strengthening.

6.2 MUD RHEOLOGY, BOTTOMHOLE PRESSURE, AND LOST-CIRCULATION PRESSURE

In order to appreciate how lost circulation can be prevented or mitigated by manipulating the drilling fluid rheology and hydraulics, let us recall how the BHP is created. During circulation, the BHP, P_w, is the sum of the static mud pressure and the frictional pressure loss in the annulus. The equivalent circulating density (ECD), ρ_{ECD}, is defined as follows:

$$\rho_{ECD} = \frac{P_w}{gD_{TVD}} = \rho_{MW} + \frac{\Delta P_{APL}}{gD_{TVD}} \qquad [6.1]$$

where ρ_{MW} is the average mud density in the annulus; ΔP_{APL} is the APL; D_{TVD} is the true vertical depth; g is the acceleration of gravity.

The average mud density in the annulus is affected by the amount of drill cuttings in the return flow. The BHP can therefore be represented as follows [7]:

$$P_w = P_{static} + P_{cuttings} + P_{dynamic} \qquad [6.2]$$

where P_{static} is the static pressure of the mud without cuttings; $P_{cuttings}$ is the hydrostatic pressure due to the cuttings suspended in the mud inside the annulus; P_{dynamic} is the dynamic pressure caused by the circulation, pressure surges and swabs during trips, etc. The static component of the BHP, P_{static}, can be evaluated if the pressure—volume—temperature data for the mud are available (the mud density may change along the well).

Eqs. [6.1] and [6.2] imply that the annulus is open to the atmosphere. We will see later in this chapter that, in some managed-pressure drilling techniques, a backpressure is applied on the surface, which will produce an extra term on the right-hand side of Eqs. [6.1] and [6.2].

The goal is to drill so that the BHP remains above the minimum allowed BHP when pulling out of the hole and remains below the maximum allowed BHP while circulating, starting pumps, or running in hole. The minimum allowed BHP is determined from the borehole stability considerations and the formation pore pressure. If the BHP drops below the pore pressure, formation fluid influx may start. If the BHP drops below the limit required for borehole stability, borehole instabilities may occur when the circulation is stopped. The maximum allowed BHP is determined by the lost-circulation pressure.

As we saw in Chapter "Mechanisms and Diagnostics of Lost Circulation," the lost-circulation pressure may depend on the pore pressure, the minimum in situ stress, rock plasticity in the near-well area, temperature changes, etc. Keeping the BHP below the lost-circulation pressure is one of the most effective preventive measures against lost circulation. According to Dupriest [8], 40% of lost-circulation events at a major operator were resolved by reducing the ECD rather than using loss prevention materials or other preventive techniques. Dupriest [8] underlined also that combatting lost circulation by ECD control is an optimization exercise since it increases the risk of well control incidents and borehole instabilities.

Significant uncertainty about the magnitude of the lost-circulation pressure may impair implementation of loss prevention measures in practice. Let us have a closer look at the lost-circulation pressure when losses are caused by different mechanisms.

If lost circulation is due to induced fractures, an essential parameter controlling the losses is the fracturing pressure. It is the wellbore pressure at which a sufficiently long fracture is induced that causes measurable losses. The fracturing pressure is, in fact, the lost-circulation pressure in this case. This parameters depends on whether there are already some small flaws/cracks on the borehole wall induced, eg, by the drill bit action. If there are no such preexisting fractures, the fracturing pressure is determined by the stress concentration around the well. If there are preexisting fractures originating from the borehole wall, the fracturing pressure will be reduced [9]. Thus, in the same formation, the fracturing pressure will be different in places with and without preexisting fractures. This can be illustrated using the pressure versus time curve obtained during extended leakoff test (XLOT) with and without a preexisting fracture at the borehole wall (Fig. 6.2).

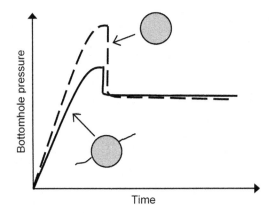

FIGURE 6.2 Pressure vs time in extended leakoff test (XLOT) with (solid line) and without (dashed line) a preexisting fracture. *Based on Lavrov A, Larsen I, Bauer A. Numerical modeling of extended leak-off test with a pre-existing fracture (in print). Rock Mechanics and Rock Engineering 2015.*

Whether with or without preexisting fractures, the fracturing pressure is a function of in situ stresses, in particular the minimum in situ stress. A variety of stress measurement methods are available in rock mechanics. These methods have been routinely used in mining or civil engineering. In those industries, a direct access to rock is often possible which enables the use of stress measurement methods based on overcoring, anelastic strain recovery, or the Kaiser effect (chapter: Stresses in Rocks) [10,11]. Also, natural fractures can often be directly observed in the outcrop rock or in the underground cavities in mines. In oil and gas industry, such direct access to rock is not available. For this reason, stress measurement techniques are most often limited to XLOT and varieties thereof [10,12,13].

The complexity of evaluating the lost-circulation pressure is illustrated in Fig. 5.24. If lost circulation is caused by fluid escaping into a natural fracture system, the lost-circulation pressure is determined by the fracture apertures, orientations, connectivity, and in situ stresses. In particular, with very wide fractures, losses may start as soon as the BHP exceeds the fluid pressure inside the fracture. If the aperture is more narrow, the fracture needs to be reopened first, which requires that the BHP exceeds the fracture reopening pressure. The latter is determined by many factors, including the fracture orientation. Since fracture orientations are often random, it makes sense to speak about a *spectrum* of fracture reopening pressures and, thus, a *spectrum* of lost-circulation pressures, rather than a single value, in such formations. The lost-circulation pressure estimated from XLOT or leakoff test (LOT) data at the casing shoe may not be representative for the interval to be drilled in this case (chapter: Mechanisms and Diagnostics of Lost Circulation).

Preventing lost circulation by keeping the BHP below the lost-circulation pressure requires that the BHP can be accurately estimated during drilling. During circulation, BHP depends on the rheological properties of the drilling mud since they determine the APL. The rheological properties are affected by the downhole temperature and pressure. They are also affected by the concentration and properties of the drill cuttings carried in the mud. These variables introduce significant uncertainty into the estimate of BHP. For this reason, it is sometimes recommended to keep the *static* BHP (ie, the weight of the mud column) below the lost-circulation pressure [14].

The goal of keeping the BHP low to avoid lost circulation can sometimes contradict the other goal, namely that of keeping the BHP high enough to avoid formation fluid influx. Since a kick is usually considered a more dangerous event than lost circulation, keeping the BHP high is often the preferred strategy. For that reason, given a narrow drilling margin, the choice can be made to proceed with losses rather than risk a kick.

Even when well-control considerations permit a reduction of BHP, often only a limited reduction can be achieved in practice by purely hydraulic means. Ideally, preventing lost circulation by ECD control could be achieved by reducing the mud weight, the mud rheology, or the pump rate. In reality, however, this may lead to other drilling problems. For instance, Eq. [6.2] suggests that BHP can be reduced by reducing the cuttings load. This, however, requires the pump rate to be increased, so that cuttings are transported faster out of the hole. Increasing the pumping rate means, however, that the dynamic component of the BHP, P_{dynamic}, increases, counteracting possible favorable effects of reducing the cuttings load. Alternatively, the cuttings pressure can be reduced by reducing the cuttings production; ie, by lowering the rate of penetration.

As another strategy, reducing BHP by reducing the static pressure (using a lighter mud) can be tried. This, however, may increase the risk of influx when the pumps are stopped, if the static pressure drops below the formation pore pressure.

Another possibility evident from Eq. [6.2] is to reduce the dynamic component of the BHP. This can be achieved either by reducing the pump rate or by lowering the high-shear rheology of the mud. Reducing pressure surges by consistent use of gradual pump start-up procedures may mitigate losses during start-ups [15,16]. During circulation, however, reduced pump rate may jeopardize the hole cleaning. Making the mud thinner at high shear rates may jeopardize the hole cleaning and promote barite sagging. Sagging results in density segregation, which in turn may promote lost circulation in the lower part of the hole. On the other hand, if the fluid is too thick and has an excessive gel strength, an excessive ECD is needed to break the gel when the pumps are started. High ECD may result in fracturing the formation and in losses. On the upside, higher yield stress (and higher rheology at low shear rates, in general) will reduce losses into natural fractures (chapter: Mechanisms and Diagnostics of Lost Circulation) [17].

Pressure variation during tripping is a common source of losses and influxes. Running pipe in hole increases the BHP and may induce losses.

The pressure surge can be mitigated by reducing the plastic viscosity, yield stress, and gel strength of the mud. Optimizing the structure and geometry of the well and the drillstring, in particular reducing constrictions in the annulus, will reduce pressure surges. Reducing the tripping speed is another practical way to limit the pressure surges [18].

Losses into natural fractures are greatly affected by the rheological properties of the drilling fluid. We saw in Chapter "Mechanisms and Diagnostics of Lost Circulation" that losses can eventually stop by themselves if the yield stress of the drilling fluid is sufficiently high. The plastic viscosity of the fluid, on the other hand, affects only the flow rate into the fracture, but not the total lost volume. The flow rate of mud losses is higher with more shear-thinning muds. Preventing/reducing losses while drilling in naturally fractured rocks involves therefore optimization of the drilling fluid rheology so as to increase its yield stress without increasing the plastic viscosity (Higher plastic viscosity would only mean higher frictional losses and thus higher ECD and risk of lost circulation.). From this viewpoint, fluids with high yield stress, such as, eg, mixed-metal hydroxide muds, are superior to fluids with high plastic viscosity, such as silicate muds [19]. Majidi et al. [20] recommended the use of drilling fluids with the yield stress greater than 10 lbf/100 ft^2 (4.8 Pa) while drilling in heavily fractured rocks (The yield stress value refers to downhole pressure and temperature.).

When preventing mud losses in naturally fractured reservoirs, one should strive to minimize the formation damage. From this viewpoint, preventing mud losses by optimizing the fluid rheology is superior to using particulates since it is easier to clean the fractures afterward in the former case.

Preventing losses when running the casing or cementing the well can be achieved by accurately designing the tripping schedules to ensure that the ECD stays below the formation breakdown pressure (FBP). It is important to avoid both the excessive running speed and excessive accelerations in order to reduce the pressure surge when running the casing [16]. Hydraulics software can be used to adjust the running speed in real time during the operation.

Continuous circulation drilling, a type of MPD covered later in this chapter, can be used to eliminate pressure surges during connections.

Preventing or reducing losses while running casing or cementing the well is possible by reducing frictional losses in the annulus. One way to achieve this is by reducing the effective viscosity of the drilling fluid before pulling out of the hole. This solution, however, can be costly to implement, and risky in deviated wells [16].

ECD control becomes crucial when drilling wells with narrow operational drilling margins; eg, deviated wells, high-pressure high-temperature (HPHT) wells, and deepwater wells in depleted reservoirs [21]. An example of using ECD control in a deepwater well is described in Box 6.1.

BOX 6.1 ECD Control During Drilling in a Depleted Reservoir in the Gulf of Mexico

A successful implementation of ECD control when drilling a deepwater well in a depleted reservoir was described by Willson et al. in their SPE paper 84266 "Wellbore stability challenges in the deep water, Gulf of Mexico: case history examples from the Pompano Field."

The well was drilled in a formation composed of depleted sands and over-pressured shales. The first attempt to drill with a mud weight of 13.1 ppg, which would prevent borehole instabilities in shales, led to total mud losses in sands.

A different strategy was chosen when drilling a sidetracked well. The well was drilled with 11.4 ppg mud weight in sands. Upon entering the shales, the mud weight was slightly increased, and the rate of penetration was controlled in such way that the ECD remained at 12.6 ppg; ie, lower than the 13.1 ppg mud weight used in the original well. The increased mud weight and ECD did not produce losses in sands, presumably because a filter cake had earlier been built there that prevented losses. This mud weight could not, however, prevent hole instability and subsequent packoffs in shale caused by pressure drop while pulling out of the hole. The hole was successfully cleaned and was drilled to the target depth. Thus, in this case, the benefits of avoiding lost circulation in sands outweighed the detrimental effects of instabilities in shales, which were handled successfully.

6.3 EFFECTIVE HOLE CLEANING

Poor hole cleaning may result in cuttings accumulation at the bottomhole assembly when pumps are shut down. Larger and denser particles settle faster because the terminal settling velocity increases with particle size and density, as we saw in Chapter "Drilling Fluid." When the pumps are turned on again, the accumulated cuttings may cause a pressure surge and subsequent mud losses. In addition, cuttings accumulation in the mud increases its rheology. This, in turn, increases the APL. Moreover, the mud weight in the annulus increases when the mud contains more cuttings in suspension. Therefore, hole cleaning is one of the most basic preventive measures against lost circulation.

Hole cleaning can be improved by [16,22]

- reducing cuttings production by lowering the rate of penetration;
- increasing the pump rate, thereby increasing the annular velocity and cuttings transport out of the hole;
- reducing the settling velocity by reducing the size of cuttings; this can be achieved, eg, by using smaller cutters;
- reducing the settling velocity by increasing the plastic viscosity of the mud; this can be achieved by using high-viscosity sweeps designed to place a high-viscosity fluid above the bottomhole assembly before a connection (effective in vertical wells);
- pipe rotation, which is particularly effective in deviated wells since it helps agitate the cuttings deposited on the lower side of the annulus.

Unfortunately, while improving the hole cleaning, some of the measures listed above may aggravate other drilling problems. For instance, increasing the pump rate may push the BHP above the fracturing pressure, thereby inducing losses rather than preventing them. The same is true about increasing the plastic viscosity: it increases the APL, which may promote lost circulation.

When preventive measures such as ECD control or improved hole cleaning cannot eliminate losses, more sophisticated techniques can be used. For instance, an attempt can be made to increase the lost-circulation pressure of the formation by wellbore strengthening.

6.4 WELLBORE STRENGTHENING AND LOSS PREVENTION MATERIALS

Wellbore strengthening is a preventive measure against lost circulation that aims to increase the lost-circulation pressure of the formation. If successful, wellbore strengthening can reportedly increase the FBP gradient by 8.0 ppg and the fracture propagation pressure (FPP) gradient by 3.0—6.0 ppg [23].

Wellbore strengthening can be achieved either by continuous circulation of particles in the mud or by squeezing a pill into the formation [24]. Hesitation squeeze has been successfully employed in formations with lower permeability [8]. Particulate materials used for wellbore strengthening are often called *wellbore strengthening material (WSM)* or *loss prevention material (LPM)*. Calcium carbonate, graphite, nut hulls, ground calcined petroleum coke, frac sand, coarse barite, sawdust, and other particulates have been used as WSM. WSM concentrations of 50—150 kg/m^3 are common [25]. Wellbore strengthening is believed to be an effective measure against lost circulation caused by natural or induced fractures, as long as the fracture width does not exceed 2 mm and the rock is sufficiently permeable. WSMs have been used with water-base and oil-base muds (OBMs) [23,24].

Reports of successful treatments with WSM in high-permeability rocks abound. Wellbore strengthening reportedly works equally well against losses caused by natural and induced fractures in such rocks [8,26]. An example of continuous application of WSM in a deepwater extended-reach well in the Gulf of Mexico was described in Ref. [22]. Losses were caused by a number of factors in that case, including poor hole cleaning, reservoir depletion, barite sagging, natural fractures, and low rock strength.

Applications of particulate WSMs in low-permeability rocks have been both success and failure [8,27]. Failures are usually attributed to the low fluid loss in such rocks, which limits the fluid filtration through the fracture wall. At the same time, special muds have been developed that overcome this difficulty and work in low-permeability rocks; eg, the "ultralow fluid loss mud" [27].

The ineffectiveness of particulate WSMs in rocks of very low permeability (shales) was recognized at the very dawn of the wellbore strengthening technology. However, as pointed out by Fuh et al. [23], the majority of losses occur in permeable rocks (eg, sandstones). "Impermeable" rocks (eg, shales) often have substantial strength. In addition, in situ stresses are often higher in such rocks. This reduces the risk of lost circulation caused by induced fractures. Natural fractures in shales exist and may cause lost circulation, particularly because preventive treatments with LPM work poorly here [23].

The mechanism and theory of wellbore strengthening have been a much debated issue in the industry for more than two decades. One explanation of wellbore strengthening assumed a mechanism referred to as "the fracture closure stress" [8]. According to this theory, injected particles prop the induced fractures at some distance from the borehole. This increases the fracture closure stress, effectively increasing the rock stress acting normal to the fracture plane. This is perceived as an increase in the fracturing pressure. If this mechanism is responsible for wellbore strengthening, the fracture should be opened as wide as possible for wellbore strengthening to work. This requires that the so-called "immobile mass,"—ie, filter cake—is built up in the fracture. The deposition of the "immobile mass" is facilitated by the fluid leakoff through the fracture walls, which requires that the rock permeability is sufficiently high.

Another explanation of the wellbore strengthening effect was based on the concept of "stress cage" [28]. According to this model, particles prop the induced fracture near its mouth and thereby act as a wedge in the borehole wall. The wedge action increases the hoop stress, creating the "stress cage." The fractures stay propped even after the wellbore pressure is reduced. If the wellbore pressure is later increased again, the elevated hoop stress prevents new fracturing and reactivation of existing fractures, effectively increasing the fracture gradient in the strengthened interval. According to the stress cage hypothesis, propping fractures of a larger aperture will enhance the strengthening effect. The fractures must therefore be short and wide, according to the stress cage concept. Six inches is sometimes quoted as the optimal fracture length for wellbore strengthening [28]. As pointed out in Ref. [29], it is not quite clear how to create so accurately dimensioned fractures in practice.

It was shown in an insightful and thorough analysis by van Oort and Razavi [30] that experimental evidence does not support the stress cage and the fracture closure stress theories. In particular, if the stress cage effect were at work, the reopening pressure of the fracture would increase after the treatment since this pressure is controlled by the stress concentration near the well. In laboratory experiments analyzed in Ref. [30], no such increase was observed. Only the FPP was found to increase in those studies.

Experimental data seems to be consistent with the third proposed mechanism of wellbore strengthening that is referred to as the *fracture propagation resistance* in Ref. [30]. According to this theory, the enhancement of the fracture gradient occurs because WSM seals the induced fracture near its tip so that the tip becomes hydraulically isolated from the rest of the fracture and from the wellbore. Therefore, the fracture can withstand higher wellbore pressures without propagating further. This is perceived as an increase in the fracturing pressure and fracture gradient.

The fracture propagation resistance mechanism is based on the theory of lost-circulation pressure proposed by Morita et al. in their seminal papers published in the 1990s [31−33]. This theory was already discussed in Chapter "Mechanisms and Diagnostics of Lost Circulation." The basic idea is illustrated again in Fig. 6.3. As mud invades the induced fracture, a filter cake is deposited at some distance from the fracture tip. This filter cake seals the fracture and protects its tip from the pressure that exists in the rest of the fracture, where the fresh drilling fluid has invaded. This is usually referred to as the tip screenout [24]. In a more permeable rock, the seal builds up easier, which explains why wellbore strengthening works better in more permeable rocks. According to the theory developed by van Oort and Razavi [30], the FPP in the presence of the filter cake and the noninvaded zone is given by:

$$\text{FPP} = \frac{\pi}{2} \frac{\sigma_{\min} - P_p}{\arcsin \frac{L_{in}}{L_{in} + L_{ni}}} + P_p \qquad [6.3]$$

where $\sigma_{\min}$ and P_p are the minimum in situ stress and the pore pressure, respectively; L_{in} and L_{ni} are the lengths of the invaded and noninvaded zones,

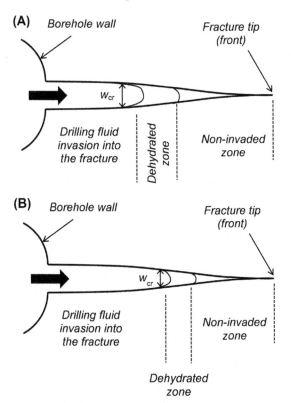

FIGURE 6.3 Schematic illustration of the effect of w_{cr} on the length of noninvaded zone. w_{cr} in (A) is greater than in (B), and thus the length of the noninvaded zone is greater in (A), too.

respectively. It is assumed in this model that the filter cake is perfectly isolating. Hence, the fluid pressure in the noninvaded zone is equal to the formation pore pressure, P_p.

The length of the noninvaded zone is on the order of [30]

$$L_{ni} = \frac{a_1 E' w_{cr}}{\sigma_{min} - P_p} \qquad [6.4]$$

where a_1 is a numerical coefficient and E' is the plane-strain Young's modulus of the rock given by

$$E' = \frac{E_d}{1 - v_d^2} \qquad [6.5]$$

where E_d and v_d are the drained Young's modulus and the drained Poisson's ratio of the rock.

w_{cr} in Eq. [6.4] is the fracture width at the sealing point. It is the maximum fracture aperture that can be sealed with the given mud in the given rock. The value of w_{cr} depends on the mud properties (fluid loss, particle size distribution and concentration of WSM) [31]. A range of w_{cr} values from 0.05 in to 0.2 in (1.3−5 mm) was reported by Fuh et al. [23]. The increase of L_{ni} with w_{cr} is easy to grasp (Fig. 6.3). The mud in Fig. 6.3A is capable of sealing a wider fracture than in Fig. 6.3B. As a result, the seal can be established farther away from the tip in Fig. 6.3A.

The length of the invaded zone in Eq. [6.3] can be evaluated from the mass balance considerations as follows [30]:

$$L_{in} = \frac{a_2 Q \sqrt{t}}{hC} \qquad [6.6]$$

where a_2 is a numerical coefficient; Q is the injection rate (the flow rate into the fracture); t is time; h is the fracture height; C is the leakoff coefficient of the formation.

The fracture propagation resistance theory is able to explain basic facts known from laboratory tests and field applications of wellbore strengthening. For instance, it explains consistently the poor performance of wellbore strengthening in low-permeability rocks. In such rocks, the leakoff coefficient, C, is relatively low, which means that the length of the invaded zone is relatively large, according to Eq. [6.6]. This reduces the FPP, according to Eq. [6.3], all the rest being equal. In a limit of an impermeable rock, $C = 0$, and FPP becomes equal to σ_{min}; ie, there is no wellbore strengthening effect.

Low permeability affects not only the FPP value, but also the way the fracture propagates. Namely, the propagation becomes more unstable as C decreases (Chapter "Mechanisms and Diagnostics of Lost Circulation"). This is intuitively clear: the fluid pumped into the fracture has nowhere else to go but through the fracture wall (leakoff) or into the newly formed fracture. If the leakoff rate is zero ($C = 0$), the injected fluid can only be accommodated by the new fracture volume. Thus, the fracture will propagate faster.

Another important implication of the fracture propagation resistance theory is the effect of w_{cr} on the FPP. The length of the noninvaded zone increases with w_{cr}. Hence, all the rest being equal, FPP increases with w_{cr}, according to Eq. [6.3] [31]. According to Eq. [6.4], the length of the noninvaded zone is greater in formations with higher Young's modulus. In such formations, the fracture aperture is smaller, and therefore the aperture equals w_{cr} farther away from the fracture tip. The FPP is thus higher in such formations. This is consistent with theoretical results of Refs. [23,26].

The length of the invaded zone has the opposite effect on the FPP. Namely, all the rest being equal, FPP decreases as L_{in} increases. Since most part of the fracture is occupied by the invaded zone, it follows that shorter fractures serve better for wellbore strengthening than long fractures. An induced fracture should therefore be sealed as quickly as possible. This is consistent with recommendations obtained with other models of wellbore strengthening, such as the stress cage model.

A wider fracture means that the seal will be established closer to the tip, given a mud with a specific value of w_{cr}. This means that wider fractures are not favored for wellbore strengthening. This conclusion of the fracture propagation resistance theory is opposite to what is advocated in other theories of wellbore strengthening. In particular, according to the stress cage theory, a wider fracture is favored since it enhances the wedge effect and thereby increases the hoop stress concentration around the well.

According to the fracture propagation resistance theory, the following requirements should be met in order to achieve good wellbore strengthening with particulate materials:

- The formation must have sufficiently high permeability (high leakoff coefficient, C). It is, however, not quite clear what the lowest permeability is where wellbore strengthening still works. Based on field tests, Fuh et al. suggested in that LPM is ineffective in rocks with permeability much smaller than 0.01 mD and is effective in rocks (sandstone) with permeability 500 to 800 mD [23]. Aston et al. report on successful wellbore strengthening in rocks with permeability 160 mD or greater [27]. These values can provide some guidance. It is, however, not clear if wellbore strengthening can be effective in rocks with permeability between 0.01 and 100 mD. A clear-cut discrimination between treatable and nontreatable permeabilities would be useful for consistent application of wellbore strengthening.
- The fracture(s) created during wellbore strengthening should be short and narrow (but wide enough to accommodate WSM particles).
- The mud must be able to seal sufficiently wide fracture apertures so as to create a sufficiently long noninvaded zone that would isolate the fracture tip.

- The seal must have sufficiently low permeability and sufficiently high mechanical strength to sustain, during subsequent drilling, the pressure difference between the noninvaded zone and the rest of the fracture. High strength also enables the seal to sustain pressure variations caused, eg, by tripping in the well. It also enables the seal to withstand an increase in the normal stress (fracture closure stress) after a squeeze treatment, when the wellbore pressure is relieved. The ability to deform reversibly without being crushed is an important property of particulate WSMs and is discussed further in this section.

In addition to the requirements in the above list, particles should have a density close to the mud density in order to stay buoyant. Furthermore, particles should not be abrasive and should have sufficient strength to endure collisions during placement [23].

If the pressure differential across the seal exceeds a certain threshold, the seal will break. This will happen at the weakest spot in the seal [26]. Laboratory tests reported in Ref. [26] suggest that the drilling fluid is then diverted toward the location of the breakthrough. The particles carried by the fluid start rebuilding the seal at that location, until the broken seal is repaired. The wellbore pressure increases again until another weak spot is destroyed, and the fluid again can rush beyond the seal. This repeated process results in a characteristic sawtooth shape of the pressure versus time curves obtained in sealing experiments in laboratory.

The performance of particulate WSMs is reportedly better in depleted formations [29]. This is in agreement with the fracture propagation resistance theory since reduced pore pressure facilitates the leakoff and, hence, the seal deposition inside the fracture.

When particulate materials are used as WSM, the term "wellbore strengthening" is somewhat misleading. The word "strengthening" implies that the rock strength (eg, tensile strength) is somehow increased, which is not the case no matter what theory is invoked to explain the improved fracture gradient. The term would be more appropriate when referring to, eg, treatments with cement or cross-linked polymers. Nonetheless, we use the term "wellbore strengthening" in this text to refer also to treatments with particulates since it is commonly used in the industry in this context.

The mechanisms of wellbore strengthening are of utmost importance for designing the treatments with particulate materials. Two main design parameters are the particle size distribution and the particle concentration (lb/bbl or kg/m^3). If the stress cage mechanism alone were responsible for the strengthening effect, particles smaller than the maximum size able to enter the fracture must have little or no effect on the treatment results. The wedge effect would be created solely by particles that are large enough. Thus, particles responsible for wellbore strengthening must have the diameter on the order of or greater than the fracture aperture at the borehole wall, according to the stress cage theory.

In the fracture propagation resistance theory, on the contrary, a broad particle size distribution is advocated since such distribution enables buildup of an effective filter cake inside the fracture. Different theories of wellbore strengthening may thus yield very different recommendations for field applications of this loss prevention method, as illustrated in Table 6.1.

As mentioned at the beginning of this section, wellbore strengthening can be applied as a continuous treatment or as a squeeze. It is pointed out in Ref. [30] that continuous treatment is more favorable than a squeeze, at least from the fracture propagation resistance theory's viewpoint. In particular, the LPM concentration required for an effective wellbore strengthening is twice as low in a continuous treatment as in a squeeze [26]. Fractures can be created in a continuous treatment without inducing losses as long as the seal forms at the very fracture mouth and prevents the fluid pressure transmission into the

TABLE 6.1 Recommended Particle Size for Wellbore Strengthening

Recommended Particle Size	Theory Behind the Recommendation	References
Equal to or larger than the fracture aperture at the borehole wall	Stress cage	[28]
Uniform (ie, narrow) particle size distribution to maximize fluid loss through the fracture wall. Particles under 100 μm are undesirable since they reduce the leakoff.	Fracture closure stress	[8]
Broad particle size distribution in order to maximize the sealing efficiency	Fracture propagation resistance	[30]
Particle size distribution should contain some percentage of particles larger than the fracture aperture. D_{90} must be approximately equal to the fracture aperture.	Laboratory experiments	[34]
Relatively uniform (ie, narrow) particle size distribution. Relatively large particles.	Laboratory experiments, field tests, theory	[23]
Broad particle size distribution: from colloidal (c. 1 μm) to c. 1 mm	Theory, field trials	[27,35]
Broad particle size distribution. Too coarse particles impair the sealing, and therefore their concentration should not exceed 5–10 lb/bbl (14.25–28.5 kg/m^3).	Laboratory experiments	[36]

induced fracture [26]. Dupriest [8] pointed out that continuous application of WSM should be used in high-permeability rocks, while hesitation squeeze is preferred in lower-permeability rocks. This suggests that squeezing is, in practice, more efficient than continuous WSM injection.

It is not improbable that more than one mechanisms are at play in wellbore strengthening. In particular, the fracture closure stress hypothesis has some common ground with the fracture propagation resistance theory. Indeed, in both models the leakoff is responsible for the buildup of the seal (also referred to as "immobile mass" [8]) inside the fracture. In the fracture closure stress model, this "immobile mass" serves to increase the fracture closure stress. In the fracture propagation resistance theory, the filter cake serves to isolate the tip from the rest of the fracture. Further experimental and theoretical research is needed in order to clarify the mechanisms behind wellbore strengthening.

The possible action of several (rather than a single) mechanisms in wellbore strengthening was implied in Ref. [27]. In the theoretical explanation provided in Ref. [27], both stress cage effect and filtration contribute to the strengthening. Aston et al. [27] point out that better strengthening is obtained with short fractures, high concentrations of LPM, and stiffer rocks (higher Young's modulus). This is similar to what the fracture propagation resistance theory predicts, too.

Experimental studies of fracture bridging and sealing suggest that the most effective way to seal a fracture is by injecting a particle mix that contains relatively large particles that can bridge the fracture near its mouth, and smaller particles that can be deposited between the large particles to create an impermeable seal [34]. As a rough estimate, particle blends with D_{90} approximately equal to the fracture aperture were found to create an effective seal. If, instead, the aperture is equal to D_{100}; ie, all particles are smaller than the fracture aperture, bridging will not occur at the mouth. Instead, particles will accumulate in the fracture farther away from the mouth and eventually bridge the fracture there. Smaller particles can then be deposited on the bridge. The accumulation of smaller particles on the bridge is controlled by the fluid loss through the fracture wall. Fine particles accumulating on the bridge can be either part of the LPM formulation, or be the ordinary barite particles present in the drilling fluid [36].

There is, unfortunately, no clear-cut rule as to what particle size is optimal for bridging. For instance, it is frequently mentioned that bridging pores is best achieved when the particle median size is equal to 1/3 of the pore median size: $D_{50}^{(particle)} \approx D_{50}^{(pore)}/3$ [37]. In the case of fractures, laboratory experiments are usually conducted with smooth fracture surfaces. Possible influence of fracture roughness on the particle bridging capacity requires further research. As we saw in Chapter "Natural Fractures in Rocks," particle retention in a fracture increases with the fracture roughness and is affected by normal and shear stresses acting on the fracture. It is therefore inherently difficult, in practice, to produce an optimal particle size distribution for in situ conditions.

The need for a certain amount of fine particles filling the space between larger particles in the seal is consistent with the experimental results obtained by Kumar and Savari [37]. In those experiments, several loss prevention materials having different particle size distribution were tested for their sealing capacity. The materials were ground marble, graphite (resilient graphitic carbon), ground nut hulls, and mixtures thereof. A fluid loaded with particles was injected into a slot. The aperture of the slot was tapered from 2500 to 1000 μm. The fluid loss through the particle seal was measured. The data reported in Kumar and Savari's Table 3 are plotted in our Fig. 6.4. It appears from Fig. 6.4 that the fluid loss somewhat decreases with decreasing D_{10}; ie, when the percentage of fines increases.

For a given particle size distribution, smaller fracture aperture results in higher differential pressure that the seal can withstand [34]. Moreover, the time needed to build the seal is shorter when the aperture is smaller. Finally, narrower fractures require smaller WSM volumes (or lower WSM concentrations) to establish the seal [34]. These experimental facts suggest that the fracture aperture is indeed the key input parameter when designing a wellbore strengthening treatment. Unfortunately, this parameter is not easy to evaluate, even for induced fractures. In a naturally fractured rock, several fractures of different orientations and possibly different genesis may be encountered along the interval to be strengthened. These fractures may have very different apertures. As mentioned previously, it rather makes sense to speak about a *spectrum* of natural fracture apertures in the given wellbore interval, instead of a single aperture value [37]. This suggests that using an LPM with a broad range of larger particles will improve bridging. The lack of accurate information about the aperture of induced and natural fractures may explain why strengthening treatments that work in one well sometimes fail in another.

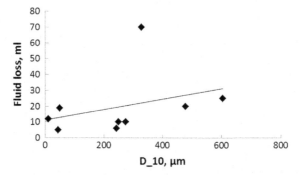

FIGURE 6.4   Fluid loss through a tapered slot vs D_{10} of LPM. *Based on the experimental data reported in Kumar A, Savari S. Lost circulation control and wellbore strengthening: looking beyond particle size distribution. AADE-11-NTCE-21 paper presented at the 2011 AADE National Technical Conference and Exhibition held at the Hilton Houston North Hotel, Houston, Texas, April 12–14, 2011. Experimental data are shown as dots.*

Particle size is not the only factor affecting the buildup of the seal. It is easy to imagine two particles that have the same size but different shapes. Will they perform equally as LPMs? Sanders et al. [36] addressed the particle shape effect in laboratory experiments. The sealing ability of differently shaped particles was measured under controlled conditions in a laboratory testing device. The fracture was modeled by two corrugated aluminum platens. An oil-base drilling fluid carrying LPM particles was injected between the platens, and the buildup of the seal was monitored. It turned out that spheroidal particles with rough surfaces were the best at building the seal. Smooth round particles (eg, proppant), flake-shaped particles, and fibers were inferior. Extensive laboratory testing of different LPMs enabled Sanders et al. to group particulates according to their sealing capacity. The following LPMs were found to have the best sealing capacity:

- sized synthetic graphite;
- graphite/coke blends;
- sized ground nut hulls;
- sized ground proprietary cellulose particles;
- ground marble (calcium carbonate).

Performance of each of these already effective materials can be enhanced by blending it with another effective material; eg, a blend of graphite and marble.

The following materials were found to perform poorly (plausible explanation for poor performance is provided in parentheses):

- sized proppant (particles are round and smooth);
- flakes (unfavorable aspect ratio);
- mica (unfavorable aspect ratio);
- fibers; eg, chopped glass fiber (unfavorable aspect ratio);
- elastomers (particles deform under pressure and thereby create flow paths through the seal).

When used in a blend with an effective LPM, such particulates do not enhance and may, in fact, impair the performance of the effective LPM.

Along with the particle shape and the particle size distribution, another important design parameter in wellbore strengthening is the LPM concentration; ie, the mass of LPM per unit volume of the fluid. High concentration enables the LPM to seal the induced fracture while the fracture is still short. Therefore, within the fracture propagation resistance paradigm, increasing the LPM concentration should enhance the FPP [23]. In a continuous LPM application, there is, however, a trade-off: making the LPM concentration too high will thicken the mud. Therefore, there is an optimal concentration of LPM. This optimal concentration depends on the rock properties, in particular the rock permeability. Fuh et al. argued that, at low concentrations (5−10% vol.), LPM will seal near the tip of an induced fracture in a permeable rock

[24]. At a medium concentration (10—30% vol.), LPM will seal at the mouth of an induced fracture, even when the rock permeability is relatively low. At high concentrations (30—50% vol.), LPM will seal near the fracture mouth, even if the rock is impermeable. In the latter case, the plug may arguably be created even though there is no leakoff through the fracture wall [24]. Laboratory experiments suggest that the minimum concentration of LPM required for fracture sealing is on the order of 10—20 lb/bbl (28.5—57 kg/m^3) [36]. The minimum required concentration depends on the fracture aperture, the particle size distribution of LPM, and the way the treatment is applied; ie, continuous or a squeeze [26]. Heavier expected losses normally call for a higher LPM concentration in the treatment; eg, 20—40 lb/bbl (57—114 kg/m^3) [36].

LPMs with lower bulk density yield more particles of a given size at the given mass concentration. This reportedly makes such LPM more effective, all the rest being equal [36].

Despite being an effective loss prevention method, wellbore strengthening is not without limitations. One that has been discussed above is poor performance in low-permeability formations. Impermeable rocks such as shale should therefore be treated with special, settable fluids rather than particulate WSMs. Such fluids, eg, cross-linked polymers or cement, can be squeezed into the formation to create a seal. This way, strengthening can be achieved even in those cases where insufficient leakoff would normally prevent wellbore strengthening with particulate LPMs [8]. A blend of a particulate with a settable material can be used as another option. The particulate material bridges the fracture, while the settable fluid, after solidification, keeps the particles in place and prevents their transport back into the well when the wellbore pressure is reduced. This provides a stable seal even when the leakoff through the fracture wall is impaired, as is the case in shale. In addition, the settable fluid reduces the permeability of the seal. The ability of the set fluid to adhere to the fracture walls is key for successful application of such systems. Other requirements include controllable and predictable setting time and stability under elevated downhole temperatures [35].

A blend of particulates with a settable fluid (cross-linked polymer) was successfully used in [35] to strengthen a 50-ft-long shale section in a well drilled with an OBM. As the particulate material, a blend of ground marble and graphite was used, with a broad size distribution of marble particles.

Poor filtration of LPM can be caused not only by low permeability of the formation, but also by the use of nonaqueous drilling fluids. Water droplets dispersed in such muds block pore throats and create damage near the fracture walls. Particulate WSMs might therefore not work very well with nonaqueous fluids [29]. Even during subsequent application of a water-based WSM pill, the fluid loss through the fracture wall may remain impaired, and the seal will not build [8]. For this reason, Dupriest [8] recommended the use of water-base muds with continuous WSM application in lost-circulation zones.

Wellbore strengthening is ineffective in vugular formations. Natural fractures having large apertures (in excess of w_{cr}) are difficult to seal, too [23].

Another problem appears when the thief zone has substantial length along the wellbore. Establishing and maintaining a seal along the entire fracture becomes more difficult as the fracture height increases [8].

LPMs are used in preventive treatments intended to increase the fracturing pressure (the fracture gradient). Their use in remedial treatments is therefore less effective [23,27]. In such treatments, fractures have already been induced while drilling with ordinary mud, and losses are occurring. Fracture faces are therefore already covered with a mud cake. An attempt to improve the fracture gradient in such a well by switching to a mud containing LPM may fail because leakoff through the fracture wall will be impaired by the already deposited mud cake. A stable impermeable seal cannot be formed when LPM is injected into the fracture in this case. LPM will therefore function only as an ordinary lost-circulation material (LCM, chapter: Curing the Losses), curing the losses that are already taking place, but not improving the fracture gradient.

Particle degradation is a concern during continuous application of WSM. Particles can be crushed as they pass through restrictions in the drill bit and in the annulus. As a results, the fine fraction of broken WSM particles (below 50...100 μm) accumulates in the mud. The accumulation of WSM fines increases the rheology of the mud and makes it heavier. For this reason, continuous application of LPM in the circulating mud is most effective when drilling relatively short sections; ie, shorter than 1000 ft [24,27]. Effective drilling fluid maintenance during continuous wellbore strengthening can be achieved, eg, by using shakers that are specifically designed to recover large WSM particles. These particles are subsequently reintroduced into the mud system [25].

It was mentioned earlier that particle systems used as LPM should be able to deform under normal stress without being crushed and should be able to regain at least part of their bulk volume when the stress is relieved. This ability to deform elastically is of utmost importance if the stress cage or fracture closure mechanisms are at work in wellbore strengthening [27]. In order to maintain the elevated hoop stresses, the seal should not be destroyed by these stresses. But even if the dominant mechanism of wellbore strengthening is the fracture propagation resistance, the LPM's ability to deform elastically is important. In particular, if the LPM is applied in a squeeze treatment, the seal must stay in place when the wellbore pressure is reduced after the squeeze. Wellbore pressure may vary during subsequent drilling. This will impose variable loads on the seal, both along the fracture (the pressure differential across the seal) and normal to fracture (opening/closing of the fracture). The ability of LPM to deform elastically will therefore improve the seal's performance as the interval is drilled further. This ability can be quantified by the LPM's *resiliency* [36,37]. Resiliency describes the ability of a particulate material to restore volume after it has been loaded and unloaded. To test for resiliency, a compressive stress of 10,000 psi is applied to an LPM sample. As a result of this loading, particles become compressed, their packing becomes tighter, and the sample contracts. The load is then relieved. The sample regains part of its initial volume. Resiliency is defined as the percentage increase in the sample's height after the load has been removed (Fig. 6.5) [37]. Some loss prevention materials have very large resiliency; eg, resilient graphitic carbon has resiliency 120%. Others, eg, ground marble, are not resilient. Experiments show that mixing a resilient material with a nonresilient one improves the resiliency of the latter [37]. If a fracture sealed with LPM undergoes opening/closing cycles during drilling, the resilient seal will be able to sustain these cycles. It will keep the fracture sealed even when the wellbore pressure increases somewhat and thus opens the fracture wider. A nonresilient material, on the other hand, will not change its size and shape as the fracture opens up. A flow path may develop around it, and the sealing effect will be lost. The pressure isolation of the fracture tip will be jeopardized. When the wellbore pressure is decreased, a resilient seal will expand. This will increase the shear stress between the seal and the fracture walls, and the seal will remain in place rather than be carried by the fluid from the fracture into the well.

   To conclude our discussion of wellbore strengthening, let us have a look at some more practical aspects. In continuous wellbore strengthening, LPM is circulated in the drilling fluid. A squeeze is a viable alternative to the continuous LPM application. In a squeeze treatment, the weak zone is strengthened by placing an LPM pill against the zone and performing a treatment that is similar to the formation integrity test (FIT): the drill bit is pulled to the top of the weak zone, the pump rate is kept constant, and the increase of the downhole pressure is monitored. The injection is stopped after the fractures have been sealed. When exactly this happens, can be estimated from the pressure versus time curve. This requires, however, that the pump rate be maintained constant

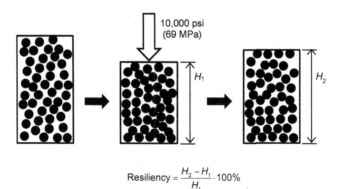

$$Resiliency = \frac{H_2 - H_1}{H_1} \cdot 100\%$$

FIGURE 6.5 Resiliency of loss prevention material.

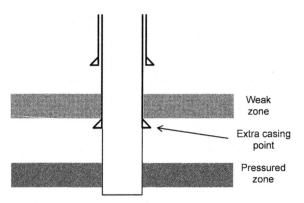

FIGURE 6.6   Drilling without wellbore strengthening: an extra casing string is required.

throughout the injection. Several squeeze cycles can be performed, with the maximum pressure increasing from cycle to cycle.

After the treatment is completed, drilling can be resumed with ordinary mud (no LPM). The mud weight can now be higher than it would be without wellbore strengthening. This may enable, eg, drilling a sequence of low-pressure weak zones and deeper high-pressure zones without setting an extra casing string. This is schematically illustrated in Fig. 6.6 and Fig. 6.7, based on a case study by Fuh et al. [24]. The well design without wellbore strengthening (Fig. 6.6) would necessitate setting an intermediate casing before drilling the pressured formation. If the casing is not set after the weak zone has been drilled, using higher mud weight when drilling the pressured zone may fracture the weak zone.

Instead of setting the extra casing, an LPM squeeze can be performed as follows. The well is drilled through the weak zone with the mud weight staying within the drilling window (Fig. 6.7A). Then, instead of setting the intermediate casing, an LPM pill is placed in the well against the weak zone (Fig. 6.7B). The pill is squeezed into the formation using one or several pumping cycles. The intention is not to exceed the FBP during the treatment. If the treatment works, the FBP will become higher than the original FBP of the weak zone. A schematic plot of pressure versus time in a successful treatment is shown in Fig. 6.8.

After the treatment is completed, drilling is resumed. The pressured zone can now be drilled with increased mud weight without risking lost circulation in the weak zone (Fig. 6.7C). A case history illustrating this scenario is provided in Box 6.2.

The detailed account of several field cases provided in Ref. [24] indicates mixed success with squeeze treatments. Continuous application provides more consistent results, but is more difficult to apply, especially in longer intervals, because solids build up in the mud over time.

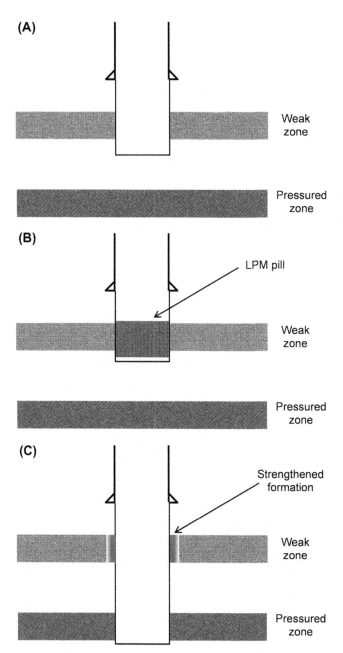

FIGURE 6.7   Drilling with wellbore strengthening: an LPM squeeze eliminates the extra casing. (A) drilling through the weak zone; (B) placing the LPM pill; (C) performing the squeeze and strengthening the formation.

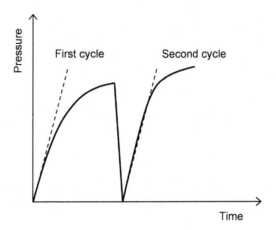

FIGURE 6.8 Pressure versus time in a squeeze treatment.

BOX 6.2 Wellbore Strengthening Eliminates an Intermediate Casing String

The following case history of successful wellbore strengthening was reported by Fuh et al. in their SPE/IADC paper 105809 "Further development, field testing, and application of the wellbore strengthening technique for drilling operations."

A vertical well was cased with surface casing at 2000 ft. Below the casing shoe, several weak zones were expected. Those zones had subnormal pore pressure (pore pressure gradient 3 to 7 ppg). The base of the weak zones was at about 6000 ft. Further down the well path, pressured zones were expected, with the pore pressure gradient of 9.5 ppg. To meet this elevated pore pressure, the mud weight had to be increased while drilling the pressured zones. This could, however, induce losses in the subnormally pressured weak zones higher up the well.

One possible solution could be to set an intermediate casing at 7870 ft; ie, after drilling through the weak zones, and continue drilling through the pressured zones with a higher mud weight.

Instead, strengthening of the weak zones was performed. After the well passed the bottom of the weak zones, an LPM pill was placed against the weak zones and squeezed into the formation. The pump rate was equal to 1/8 bbl/min during the treatment. Two pressure cycles were performed, and the pressure was increasing in both cycles. The section was then successfully drilled to the target depth with 10 ppg mud weight.

The LPM used in this treatment was a mixture of graphitized coke and calcium carbonate. In addition, barite was added to the OBM to weight it to 9.6 ppg.

This case history demonstrated that an increase in the upper mud weight limit by several ppg is possible by wellbore strengthening in a potential thief zone.

> **BOX 6.3 Wellbore Strengthening With a Settable Material**
>
> An example of successful application of a settable material to strengthen a wellbore was described by Traugbott et al. in their article "Increasing the wellbore pressure containment in Gulf of Mexico HPHT wells" (*SPE Drilling & Completion, 2007, 22,* 16–25).
>
> The 852-ft-long open hole section of an HPHT well in the western Gulf of Mexico had several weak zones causing losses. Because of these thief zones, the mud weight had to be maintained below 18.2 lb/gal. This was too low to drill the production zone further in the hole. An attempt was made to increase the upper mud weight limit by pumping LPM pills (blends of calcium carbonate, fibers, and resilient graphite). Five LPM pills were tried, with no success at increasing the upper pressure bound. The failure was attributed by Traugott et al. to fractures having very wide apertures, which disabled particle bridging. A settable material was then squeezed into the formation. This increased the maximum operational mud weight to 18.8 lb/gal. This strengthening effect was, however, not sufficient to drill the production zone. A second treatment with the same sealant increased the maximum operational mud weight to 19.1 lb/gal. This was sufficient to drill the production zone. The failure of the first treatment was attributed by Traugbott et al. to insufficient pumped volume of the settable material.

In formations with wide natural fractures, where both continuous and squeeze treatments with particulates fail, settable squeeze may be the solution (Box 6.3).

6.5 CASING WHILE DRILLING

CWD is a drilling technology in which a rotating casing is used as the drill pipe. The casing advances together with the drill bit as drilling proceeds. The drilling fluid flows down inside the casing and returns up the annulus between the casing and the formation. When the drill bit reaches the target depth, the casing is already in place. This eliminates an important source of lost circulation and well control incidents, namely tripping of the drill pipe [38]. It also has a positive effect on the well cost since the tripping time is eliminated in CWD.

CWD is effective against lost circulation in wells with narrow drilling window, such as deepwater wells. CWD is considered a viable option against mud losses caused by natural or induced fractures, including fractures wider than 2 mm [39]. Even when severe losses occur during CWD, they tend to decrease as the well is drilled ahead [40]. This reduces nonproductive time that otherwise would be spent on fighting the losses. The overall drilling time is thereby reduced.

CWD is believed to increase the fracture gradient by the so-called *plastering effect*, also called *smear effect*. The casing action grinds particles and generates a wide particle size distribution in the mud between the casing and the formation. These particles (plus LPM, if it is used) are continuously smeared into the

---

BOX 6.4 Casing While Drilling: An Onshore Well in Texas

The following case history from the Lobo field in Texas was presented by Tessari et al. in their SPE paper 101819 "Drilling with casing reduces cost and risk."

Lost circulation problems were common in the Lobo field. A decision was made to drill a well with CWD. Three offset wells drilled previously with conventional technology in the same area had severe losses, necessitating sidetracking in one of the cases. In the least troublesome of the three offset wells, losses occurred when the mud weight was 8.7 ppg. Losses could not be cured with LCM, and the weak zone had to be cemented. In order to reach the target depth of 8000 ft, an intermediate casing was set at 6917 ft. The drilling time for this least troublesome conventional well was 19 days.

The well drilled with CWD experienced losses while drilling through the weak zone. Losses decreased, however, as the well advanced. The rest of the well was drilled to the casing point at 8103 ft without losses. Mud weight up to 10.5 ppg was used during this operation. No intermediate casing was required, and the drilling time of the CWD well was 10 days; ie, half of that spent on the least troublesome conventional well.

---

borehole wall by the rotating casing. The particles seal fractures on the borehole wall and thereby increase the maximum pressure the well can sustain [30]. Casing also might polish the borehole wall and thereby eliminate small cracks.

CWD has been used to drill both onshore wells and parts of offshore wells, both vertical and directional [40]. CWD has also been used in combination with UBD [41]. CWD facilitates underbalanced operations by eliminating trips. CWD is especially beneficial when drilling depleted reservoirs and weak low-pressure zones. By reducing mud losses, CWD mitigates formation damage in reservoir sections. CWD may also eliminate casing points that otherwise would be necessary in order to avoid losses and influxes. An example of successful application of CWD in an onshore well is provided in Box 6.4.

6.6 MANAGED PRESSURE DRILLING

MPD is an effective preventive measure against lost circulation in formations with narrow drilling window. It is also a viable option in vugular formations, where losses are notoriously difficult to prevent or cure.

MPD is a collective term used for a variety of techniques that enable the annular pressure to stay within the drilling window along the well. The following drilling methods are examples of MPD:

- continuous circulation drilling;
- constant bottomhole pressure drilling;
- pressurized mud-cap drilling;
- dual-gradient drilling;
- ECD reduction tools.

In this section, we will discuss some MPD methods. Many more varieties of MPD exist than we cover here. It is not the purpose of this section to provide an exhaustive review of MPD, but rather to illustrate its basic ideas, and show how and why MPD can prevent lost circulation. The reader is referred to specialized texts on MPD, eg, Ref. [18], for a more detailed treatment of the subject.

MPD can be implemented by controlling the following parameters (or combinations thereof) [42]:

- back pressure;
- rheology and density of the drilling fluid;
- annular fluid level;
- circulating friction;
- borehole geometry.

The ultimate goal is to keep the BHP within the drilling window at all times. In particular, in order to prevent lost circulation, the BHP must stay below the lost-circulation pressure.

Continuous circulation system (CCS) is an MPD technique designed to eliminate pressure surges during connections. This is achieved by making connections without stopping circulation. The main piece of equipment in CCS is a device called *coupler*. The coupler serves as a pressure chamber installed on the rig floor above the rotary table. In the system design described in [43], the coupler is essentially a modified blowout preventer comprising two sections, the upper and the lower. During connection, continuous circulation is maintained through the lower section that is connected to the drillstring at all times. The new joint of the drillpipe is run into the upper section of the coupler. After connection, the seal between the two sections in the coupler is opened, and circulation is reestablished through the drillstring.

CCS eliminates at least one common source of lost circulation; ie, the pressure surge that would normally accompany the circulation restart after connection. Other benefits of CCS include reduced risk of stuck pipe and reduced total connection time. Applications of CCS are found in extended-reach drilling (ERD) wells, deepwater wells, formations with narrow drilling window, etc. In particular, in ERD and horizontal wells, continuous circulation prevents cuttings accumulation in the lower part of the annulus [43]. The result is better hole cleaning which mitigates the risk of lost circulation in such wells.

Constant bottomhole pressure (CBP) drilling finds applications in wells with a narrow drilling window. Without MPD, maintaining the BHP within the narrow window during drilling would require setting additional casing points. When CBP drilling is employed, the static BHP (ie, the mud weight) is close to the lower bound of the drilling pressure window. This reduces the risk of mud losses and differential sticking and improves the rate of penetration. During circulation, the BHP is higher than the static pressure because of the APL. This is true for both conventional drilling and CBP drilling. In CBP drilling, however, an additional contribution to the BHP is introduced, namely, a surface backpressure applied on the annulus. Eq. [6.2] thus needs to be modified as follows:

$$P_w = P_{static} + P_{cuttings} + P_{dynamic} + P_{bp} \qquad [6.7]$$

where P_{bp} is the backpressure. The surface backpressure can be controlled with a choke. This requires installation of extra equipment at the rig, in particular a rotating control device (RCD) and a choke manifold [18]. Backpressure control transforms the open circulation system of conventional drilling into a closed circulation system since the return line is no longer opened to the atmosphere [18].

Control of backpressure is needed, eg, in order to compensate for rapid pressure changes when the pump is turned on or off, or when tripping in and out of the hole. When the pumps are shut down, the dynamic pressure [$P_{dynamic}$ in Eq. [6.7]] drops to zero. Closing the choke as this happens increases the backpressure, P_{bp}, so that the sum on the right-hand side of Eq. [6.7] remains constant. Similarly, when the pumps are turned on, the choke is opened to reduce the backpressure and thereby prevent a surge in the BHP. Ideally, it should always be possible to maintain the BHP within the drilling window during pump start-ups and shutdowns with this technique.

If formations with different pore pressure gradients and fracture gradients are exposed in the open hole, the drilling window may become prohibitively narrow. The CBP technique enables drilling such formations by manipulating the backpressure as follows: When pumps are off, the backpressure is maintained sufficiently high to prevent formation fluid influx and borehole instabilities along the open hole. When the pumps are on, the backpressure and the pump rate are kept sufficiently low to avoid lost circulation.

In addition to the benefits of CBP outlined above, this technique enables formation integrity testing while drilling. As discussed in Chapter "Stresses in Rocks," LOTs, XLOTs, and FITs are typically performed at the casing shoe. In practice, the fracture gradient at the casing shoe may not be representative for all the formations to be drilled in the next interval. When drilling with CBP, an FIT can be performed at different locations in the interval while drilling ahead, by increasing the backpressure and monitoring flow in and out of the well. The results of such tests can be used to adjust the mud weight and other drilling parameters in order to avoid losses [18].

The benefits of using CBP drilling should be weighed against its disadvantages. In particular, CBP, similar to some other MPD techniques, requires additional equipment and modifications to the rig. The personnel must be trained in using this drilling method. Drilling an interval with CBP should be planned in advance. Moreover, CBP, like any other technology, has its limitations. For instance, maintaining CBP may still be a challenge in unconsolidated sands where packoffs of the drillstring are common [44]. Furthermore, rapid pressure changes, eg, caused by a pump failure, may challenge the choke's capacity to respond quickly. Automated pressure control systems using a dedicated backpressure pump have been developed to ensure that the BHP stays within the drilling window during rapid pressure changes [18].

Pressurized mud cap drilling (PMCD) and *floating mud cap drilling* (FMCD) are suitable methods for dealing with severe or total losses in heavily fractured or vugular formations. Such losses may lead to disappearance of the entire mud column into the formation. If this happens during conventional drilling, formation fluids such as gas can enter the well in its upper part, causing a kick.

In FMCD, a column of heavy mud is kept in the annulus. The mud level in the annulus "floats" so as to balance the pore pressure at the bottomhole. There is thus no fluid in the upper part of the annulus. Sacrificial fluid (fresh water, seawater) is pumped down the drillstring. The sacrificial fluid does not return to surface. Instead, the sacrificial fluid and the cuttings carried with it are forced into the fractures and vugs. Drilling is thus with no returns to surface.

PMCD is similar to FMCD. However, in PMCD, the annulus is completely filled with a relatively light mud. The mud is pumped into the annulus at a relatively low rate. In order to balance the formation pore pressure, a surface pressure is applied on the annulus. For instance, if the mud weight is such that the static BHP is by 100 psi (0.69 MPa) lower than the formation pore pressure, a backpressure of 100 psi (0.69 MPa) needs to be maintained at the surface. Sacrificial fluid is pumped down the drillstring.

Mud cap drilling requires large amounts of sacrificial fluid which may limit its application in some areas. Another drawback of mud cap drilling is formation damage. Furthermore, implementation of PMCD requires installation of an RCD to seal the well and pump mud into the annulus. Tripping is another challenge in mud cap drilling. Preventing migration of formation fluids into the annulus while pulling out of the hole may require prohibitively high mud injection rate into the annulus [45]. Implementation of PMCD in offshore wells may challenge the riser's pressure rating. The maximum achievable surface pressure will be determined by the riser's slip joint pressure limit [44].

When considering mud cap drilling as an alternative to conventional drilling, the above drawbacks should be weighed against benefits of this technique. In particular, PMCD and FMCD may enable drilling formations that otherwise would be impossible to drill. Various flavors of mud cap drilling have been successfully used to drill onshore as well as offshore wells [46].

An example of offshore application of PMCD is described in Ref. [45]. The well was drilled in a carbonate formation that was prone to total losses caused by karst and fractures. 268 ft (82 m) of the well was drilled with PMCD. Seawater was injected into the drillstring with a rate of 4 bbl/min (0.64 m³/min). Mud was injected into the annulus with a rate of 1 bbl/min (0.16 m³/min). The relatively high injection rate into the annulus was attributed by Terwogt et al. [45] to the large size of vugs and fractures causing losses.

Dual-gradient drilling refers to using two or more pressure gradients in the same well. This is an effective technique, in particular, when drilling formations with rapidly increasing pore pressure and/or low overburden stress. One of the main application areas is in deepwater drilling. In deepwater wells, the fracture gradient in upper sediments is often close to the hydrostatic pressure of the seawater. In such case, the drilling margin is reduced to almost zero [18]. This may require four to six extra casing strings in such wells [46]. The problem is illustrated in Fig. 5.16.

Various techniques are available for setting up a dual-gradient pressure profile along the well. In onshore wells, for instance, a lightweight fluid (eg, air or light liquid) can be injected at the point above which the gradient must be reduced. In offshore drilling, the "pump and dump" method is probably the simplest way to establish dual gradient. In this technique, mud is not returned to the rig but instead is dumped on the seafloor. As a result, the annular pressure at the mud line is equal to the hydrostatic pressure of the seawater. This is illustrated in Fig. 6.9. In the situation shown in Fig. 6.9, overbalanced drilling with a single mud weight would be impossible since it would bring the BHP above the fracturing pressure in the upper part of the interval. Dual gradient enables drilling this interval without setting extra casing points.

Environmental issues associated with the "pump and dump" technique call for more sophisticated dual-gradient methods for offshore drilling. One such method makes use of a subsea mud lift pump. The pump returns the mud to the rig using a special return line while the riser is filled with a light fluid, with the density similar to the seawater [18]. The pump effectively provides a

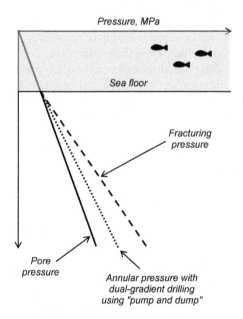

FIGURE 6.9   The principle of offshore dual-gradient drilling using "pump and dump" technique. The slope of the pressure profile changes at the seafloor.

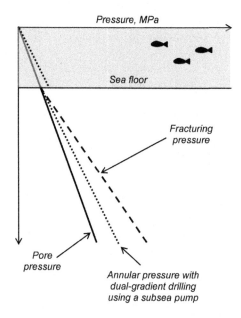

FIGURE 6.10   The principle of offshore dual-gradient drilling using a subsea mud lift pump. The pump provides a pressure boost at the seafloor.

pressure boost on the return line. This is evident as a discontinuity in the dotted line in Fig. 6.10. The pressure gradient (the slope of the dotted line in Fig. 6.10) is the same above and below the pump. The pressure boost means, however, that the pressure profile below the seafloor can now stay within the drilling window (compare Fig. 6.10 and Fig. 5.16).

The ability to follow the drilling window precisely is the main advantage of the dual-gradient method. This is especially beneficial in deepwater wells since, as we saw in Chapter "Mechanisms and Diagnostics of Lost Circulation," the drilling window is narrower in such wells, and the dual-gradient method enables drilling with fewer casing points. One of the disadvantages of the dual-gradient method is the need for extra equipment (eg, a subsea mud lift pump).

Reducing the ECD while drilling can be achieved by using an ECD reduction tool. The tool creates a pressure boost in the annulus and thereby reduces the ECD below the location of the tool [18].

The tool essentially consists of a turbine motor (driven by the circulating fluid), a pump, and the seals used to ensure that all fluid flows through the tool. The tool has been tested in onshore and offshore wells. A crusher needs to be installed upstream of the pump if large drill cuttings are present in the mud.

The ECD reduction tool has been advertised as a low-cost alternative to more advanced MPD techniques, such as the dual-gradient drilling. The disadvantage of using the ECD reduction tool is its poor performance during tripping: Acting as a constriction in the annulus, it increases the annular pressure variations [18].

6.7 AIR DRILLING AND UNDERBALANCED DRILLING

Air drilling refers to a collection of drilling methods where a gas or a liquid—gas system is used as the drilling fluid. Air drilling includes:

- dry air drilling (dust drilling);
- mist drilling;
- foam drilling;
- aerated drilling.

The percentage of gas in the drilling fluid determines the specific weight of the fluid. The gas percentage decreases from dust drilling to aerated drilling. Very low mud weight achieved with air drilling brings about several advantages compared with the conventional, all-liquid drilling:

- reduced risk of differential sticking;
- improved rate of penetration;
- reduced formation damage;
- improved formation evaluation.

The rate of penetration, in general, increases from aerated drilling toward dust drilling. The cuttings-carrying capacity increases from dry air to aerated drilling, which means that the annular velocity required for cuttings transport is highest with dry air.

Drawbacks of air drilling include extra costs of equipment and materials, for instance, the air compressors and the de-aerator tank in aerated drilling, or large pits for the foam and the extra chemicals used to break the foam in foam drilling. Modifications to the rig are necessary, for instance, installation of a rotating diverter and a four-phase separator in aerated drilling [47]. Moreover, application of measurement while drilling with air drilling calls for alternative signal transmission techniques; eg, electromagnetic signal transmission with dry air drilling, instead of mud pulse telemetry used with all-liquid muds [38,48].

Main application areas of air drilling are within

- hard rock drilling;
- drilling through lost-circulation zones;
- drilling in depleted formations;
- drilling through pay zones, where it is important to minimize the formation damage.

In particular, air drilling has been used to prevent losses in unconsolidated shallow formations (sand, gravel) [49].

In *dry air drilling*, also known as *dust drilling*, the compressed gas (air) is the only constituent of the drilling fluid. This makes dust drilling the cheapest type of air drilling by reducing, in particular, the costs associated with handling the chemicals. Small amounts of hammer oil can be injected into the well to provide some lubrication and cooling and reduce corrosion. Due to the lowest specific weight of the fluid, dust drilling can achieve the highest rate of penetration. High annular velocity in dry air drilling improves the hole cleaning. The downside, however, is the increased mechanical force exerted on the bottom hole assembly. In the absence of a liquid mud column, downhole shock and vibration may become an issue. Successful application of dust drilling requires that there be no formation fluid influx into the well. This limits the application of this technique to dry formations. Once influx occurs, mist or foam drilling can be employed. The risk of downhole fires is another drawback of dust drilling.

Dry air has been successfully used in onshore drilling. In particular, dry air is often used to drill top sections in the Northeast United States. Air flow rates between 2000 and 6000 scfm are common. In 2014, out of 111 rigs drilling in the Marcellus shale, 23 were drilling with air [48].

Mist drilling employs a gas (air), with some added liquid and surfactants, as the drilling fluid. The rate of penetration is between the dust drilling and the aerated drilling. Mist drilling reduces the risk of downhole fires and allows minor influx of formation fluids into the well.

Foam drilling makes use of a stiff foam as drilling fluid. Foam is created by mixing air, water, and surfactants.

Aerated drilling is performed with a two-phase mixture of water and air (or nitrogen). Similar to other air-drilling fluids, aerated mud has a reduced density compared with conventional all-liquid drilling.

Depending on the air-to-mud ratio, air drilling can be performed with the BHP below, equal to, or slightly higher than the formation pore pressure. Drilling with the BHP below the formation pore pressure is known as *underbalanced drilling (UBD)* and is a very effective way to prevent lost circulation. In addition, UBD improves the rate of penetration. It also reduces the formation damage, by preventing the mud from flowing into fractures and pores, and by reducing the exposure time of the reservoir to the drilling fluid [4]. At the same time, UBD presents some unique challenges. In particular, it may induce sand production in unconsolidated sands [47]. In shale, UBD may provoke instabilities and hole collapse [50].

Operators have had varying experience with aerated muds. According to some reports, aerated muds promote corrosion, erosion of tubulars, and poor hole cleaning [51]. On the other hand, in geothermal wells in New Zealand aerated drilling was found to offer a viable solution in shallow formations where normally mud losses would occur with overbalanced drilling [52]. Polycrystalline diamond compact drill bits were found to counter the problem of increased wear of roller cone bits.

Aerated drilling is particularly beneficial while drilling the last stages of geothermal wells, where the rock is permeable due to fractures, and severe or total losses would occur with all-liquid drilling fluids [53]. In addition to eliminating mud losses, aerated drilling fluids reduce the formation cooling which is important in view of subsequent use of the reservoir for geothermal energy production. An example of using UBD in geothermal drilling is provided in Box 6.5.

An example of a geothermal well drilled with progressively lighter drilling fluids was described in Ref. [54]. The casing program of this vertical well comprised four sections. Section 0 (surface casing, from 0 to 90 m), Section 1 (anchor casing, from 90 to 300 m), and Section 2 (production casing, from 300 to 800 m) were drilled with mud or water. Section 3 (production liner, from 800 to 2235 m) was started with water and then switched to aerated water, after severe losses occurred. Drilling continued with aerated water until the well reached the target depth.

The use of aerated fluids in offshore drilling for oil and gas has been limited. In onshore drilling for oil and gas, aerated drilling has been used successfully to mitigate lost-circulation issues. Aerated drilling of onshore wells in Abu Dhabi was reported in Ref. [55]. Lost circulation problems were experienced when drilling 17-1/2″ sections through low-pressure aquifers. Before deployment of aerated drilling, reducing the density of the water-base fluid by adding crude oil was tried. The resulting high waste management costs motivated implementation of aerated drilling. After initial experience was

---

BOX 6.5 Underbalanced Drilling of a Geothermal Well in El Salvador

The following case history of using aerated drilling was presented by S.A. Ramos at the Short Course on Geothermal Development and Geothermal Wells organized by UNU-GTP and LaGeo, in Santa Tecla, El Salvador, March 11–17, 2012.

Aerated drilling was employed while drilling the final stage of a deviated well in the Chinameca Geothermal Field in El Salvador in 2009. The target depth of the well was 1720 m. The well was drilled with a water-base mud. The mud composition included lignites, lignosulfonates, sodium carboxymethylcellulose, and partially hydrolyzed polyacryl amide. The last, 8-1/2″ stage of the well was started at 1080 m, after setting the 9-5/8″ casing.

Losses of 10–20 m³/h occurred at 1248 m. Drilling continued with increasingly more severe losses up until 1267 m, at which point total losses occurred. The decision was made to switch to an aerated drilling fluid.

A mixture of water and air was tried first, with still no returns. A foaming agent was then added to the fluid. This restored the circulation. The water–air foam was generated with the following composition: water flow rate 250 to 350 gpm, air flow rate 700–1700 cfm, foaming agent volume concentration 0.05%–0.20%. Drilling continued successfully to the target depth.

gained with this technique, the drilling costs (including waste disposal costs) were found to be lower by 17% compared with all-liquid drilling.

6.8 EXPANDABLE TUBULARS

One of the most straightforward ways to prevent lost circulation would be to set an extra casing string and adjust the mud weight in the next interval to keep it within the drilling margin. However, each additional casing string reduces the wellbore size. The expandable openhole drill liner technology enables setting extra casing without reducing the wellbore diameter [56]. In this technology, a drill liner is run in hole and is expanded by moving an expansion cone up along the well (Fig. 6.11). The cone is moved by two forces: the pressure difference (higher pressure below the cone than above it) and the pulling action of the drill pipe [38]. The vertical force acting on the liner during expansion is lower than the buoyant weight of the liner. The expansion force created by the cone leads to irreversible (plastic) increase of the liner's diameter, typically on the order of 6−16% [57]. The top of the expandable liner is coated with an elastomer that creates a seal upon expansion.

The first commercial application of expandable tubulars was in the Gulf of Mexico in 1999 [57]. Since then, many success stories have been reported, both with and without cementing. With expandable tubular technology, cementing can be performed either before or after the expansion. Cementing prior to the expansion requires that cement remains liquid until the expansion

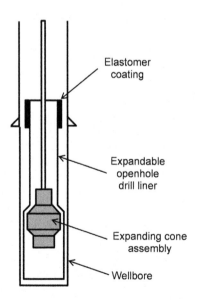

FIGURE 6.11 The principle of expandable tubular technology.

is completed. During expansion, the liner squeezes the cement up the annulus. The resulting cemented annulus is narrower than what could be achieved in a regular cementing job. (Pumping cement into such a narrow annulus would normally break the formation.)

The primary applications of expandable tubular technology are for improving the well architecture, improving the completion design, and isolating unstable formations and lost-circulation zones [58].

6.9 PREVENTING LOST CIRCULATION DURING WELL CEMENTING

Normally, lost circulation should be cured and loss zones should be sealed before the cementing job begins. Sealing loss zones during drilling considerably reduces the risk of lost circulation during cementing. However, the rheology and gravity of fluids pumped during a well cementing job are usually greater than those of the drilling mud. Therefore, there is some extra risk of fracturing the formation during cementing, even if there were no losses when the well was drilled. To avoid this risk, the downhole pressure during cementing should be kept below the upper bound of the operational pressure window, just like during drilling. Reducing the ECD is therefore the primary method of preventing lost circulation during well cementing. An obvious way to prevent losses in this case is achieved by reducing the flow rate and thus the APLs. However, the flow rate during a cementing job must be sufficiently high to effectively displace the mud from the annulus in order to ensure high-quality cementing.

Lost circulation during cementing jobs can be prevented or mitigated by using a lightweight cement since this reduces the hydrostatic pressure in the annulus. However, reducing cement density by increasing the water—cement ratio of the cement paste decreases compressive strength and increases permeability and porosity of set cement. To resolve this problem, special formulations of lightweight cement have been proposed, manipulating, for instance, the particle size distribution of the cement powder. To this end, the cement powder with a trimodal particle size distribution (ie, containing particles of roughly three sizes) and loaded with fibers has been used as a lightweight cement [59]. Adding extenders to well cement is another way of reducing its density [49]. The density of the slurry can be significantly reduced (down to $900 \, kg/m^3$ or even lower) by adding hollow microspheres to the formulation (Box 6.6) [60]. The use of foamed cement is an effective preventive measure against lost circulation in cementing jobs, because of the low density of such cements.

ECD during a well cementing job can be reduced by lowering the plastic viscosity and yield stress of the mud circulated prior to the cement placement, thereby reducing the frictional losses during cementing. Another technique for reducing the ECD involves application of a lightweight fluid or a Newtonian

BOX 6.6 Primary Cementing With Ultralightweight Cement in Fractured Carbonates

The following example of using ultralightweight cement for primary cementing in water-bearing fractured carbonates in the Middle East was described by El-Hassan et al. in their SPE paper 92374 "Using a combination of cement systems to defeat severe lost-circulation zones."

The water-bearing carbonates were drilled using a 9.1 ppg drilling fluid, with partial or severe losses during drilling. The conventional practice during casing cementing would then be to inject two cement slurries: a lightweight lead slurry of 10 ppg and a tail slurry of 15.8 ppg. Such two-slurry cementing often resulted in severe or total losses during cementing jobs.

To combat lost circulation during primary cementing, a three-slurry cementing schedule was introduced. The first slurry was an ultralightweight cement (8.95 ppg). The second slurry was a lightweight cement slurry (10 ppg). The third slurry was a conventional cement tail slurry of 15.8 ppg.

The three-slurry schedule was applied to cement the 13-3/8″ casing in the 16″ well that had experienced partial losses (100 bbl/h, ie, 16 m³/h) during drilling. 440 bbl of the first slurry, 580 bbl of the second, and 95 bbl of the third were pumped up the annulus. Returns were never lost during this job. This proved the viability of the ultralightweight cement as part of the three-slurry system in water-bearing carbonates.

When severe losses are observed during drilling, total losses may occur during cementing. El-Hassan et al. recommended adding fibers to the ultralightweight slurry in such cases, to seal the loss zone effectively at the beginning of the cementing job.

fluid ("preflush") before injecting spacer and cement. In a deviated well, this strategy may, however, lead to the lightweight fluid ending up in the less deviated part of the well, thereby reducing the hydrostatic pressure in the lower part of the annulus. This may lead to well control issues at the end of the cement job by allowing formation fluid influx [61]. Moreover, Newtonian fluids, having relatively low viscosity compared with the mud in place can channel through the mud and thereby create a preferential flow path for the spacer and cement injected afterward. This can result in poor cementing quality because part of the mud is left undisplaced, and part of the annulus is left uncemented.

In addition to keeping the BHP below the lost-circulation pressure, some other preventive measures against lost circulation during cementing are available. For instance, a spacer containing a plugging material can be pumped before cement, LCM can be added to the cement slurry itself, or a thixotropic cement can be used [49].

Computer simulations and detailed planning of well cementing jobs are among the most important preventive measures against lost circulation in well cementing.

6.10 SUMMARY

We have seen in this chapter that a variety of loss prevention techniques are available. Each of these techniques has its advantages and disadvantages. Therefore, there is no single, universal loss prevention technique. Rather, the choice between the techniques should be based on their performance in different formations. It should also take into account the mechanisms of lost circulation in the specific wells (eg, a vugular rock, a formation with low fracture gradient reduced by depletion, etc.). Some examples of preventive measures against losses of different severity and caused by different mechanisms are provided in Tables 6.2 and 6.3, respectively [39].

Keeping the BHP below the lost-circulation pressure is the most efficient way to prevent losses. In conventional drilling, this can be achieved by avoiding constrictions in the annulus, by keeping the hole clean, by reducing the rheology and gravity of the fluid, or by reducing the pump rate. These measures, however, often contradict each other, for instance, reducing the pump rate without reducing the rate of penetration may compromise the hole cleaning.

Wellbore strengthening can be used as an effective preventive measure against mud losses in depleted formations, deviated, and horizontal wells. Wellbore strengthening, if successful, increases the lost-circulation pressure of the formation so that the well can be drilled further with a higher mud weight

TABLE 6.2 Examples of Preventive Measures According to the Severity of Expected Losses

Severity of Losses	Preventive Measures
Seepage (<10 bbl/h)	• Keeping the BHP below the lost-circulation pressure • Wellbore strengthening • MPD • CWD
Partial (10–100 bbl/h)	• Keeping the BHP below the lost-circulation pressure • Wellbore strengthening • MPD • CWD • Solid expandable systems
Severe (>100 bbl/h)	• Keeping the BHP below the lost-circulation pressure • MPD • CWD • Solid expandable systems
Total (no returns)	• Keeping the BHP below the lost-circulation pressure • MPD • CWD • Solid expandable systems

TABLE 6.3 Examples of Preventive Measures According to the Mechanism of Expected Losses

Mechanism of Lost Circulation	Preventive Measures
Losses into porous matrix	• Keeping the BHP below the lost-circulation pressure • Sized particles • MPD
Losses in vugular or cavernous formations	• Keeping the BHP below the lost-circulation pressure • MPD (PMCD) • UBD • CWD
Losses into natural fractures	• Keeping the BHP below the lost-circulation pressure • Optimizing the drilling fluid rheology: Increase the yield stress of the fluid (works for fractures up to a certain aperture) • Wellbore strengthening (works for fractures up to c. 2 mm wide) • CWD (may work also for fractures wider than 2 mm) • MPD • UBD
Losses caused by induced fracturing	• Keeping the BHP below the lost-circulation pressure • Wellbore strengthening • CWD • MPD • UBD

without setting extra casing. Wellbore strengthening may also mitigate the risk of lost circulation in subsequent cementing.

Two types of materials are currently used for wellbore strengthening:

• particulates mixed into the drilling fluid;
• settable systems (eg, cement, cross-linked polymer systems, etc.)

Particulates (eg, the so-called high-solids, high-fluid-loss systems) are most effective in permeable rocks since they require fluid leakoff through the fracture wall to create a seal. Such systems can be either circulated continuously as the thief zone is being drilled through, or be applied as a squeeze or a hesitation squeeze. Continuous application of LPM provides more consistent results, but is more difficult to apply, especially in longer intervals, because solids build up in the mud over time. Laboratory and field data suggest that the particle size distribution should be such that larger particles establish a bridge in the fracture, on which finer particles are then deposited. This should ensure low permeability of the seal and effectively isolate the fracture tip from the wellbore. The seal should be established near the fracture mouth. The fractures induced during the treatment should be relatively short. Designing the particle size distribution requires some information about fracture apertures. This information is rarely available. This represents one of the main challenges in application of particulate materials for preventing and, as we will see in the next chapter, for curing mud losses.

Treatments with settable materials can be a viable preventive measure against lost circulation in low-permeable and impermeable rocks (eg, shale). The treatment is performed as a squeeze. Seals established by settable materials may substantially impair the permeability of the formation.

MPD provides an effective solution for wells with narrow drilling margins; eg, deepwater wells and wells in depleted reservoirs. It also may help preventing severe or total losses that otherwise would occur in vugular or heavily fractured formations. The main disadvantages of MPD are the need for extra equipment, modifications to the rig, and personnel training. The crew experience and commitment are vital for successful deployment of MPD. Moreover, MPD is not a cure-all solution against lost circulation since pressure surges during trips still occur in most MPD techniques. Finally, most of the MPD techniques still need to reach the degree of maturity that will make them a widely accepted strategy for preventing lost circulation.

When lost circulation cannot be prevented, the second best option is to plan for lost circulation when designing the drilling program. Having the right materials in sufficient quantities ready at the rig will allow a timely and effective treatment when losses occur. Treatments used for curing different kinds of losses are the subject of the next chapter.

REFERENCES

[1] Ameen MS. Fracture and in-situ stress patterns and impact on performance in the Khuff structural prospects, eastern offshore Saudi Arabia. Mar Pet Geol 2014;50:166−84.

[2] Straub A, Krückel U, Gros Y. Borehole electrical imaging and structural analysis in a granitic environment. Geophys J Int 1991;106:635−46.

[3] Kristiansen TG, Mandziuch K, Heavey P, Kol H. Minimizing drilling risk in extended-reach wells at Valhall using geomechanics, geoscience and 3D visualization technology. SPE/IADC paper 52863 presented at the 1999 SPE/IADC Drilling Conference held in Amsterdam, Holland; March 9−11, 1999.

[4] Abdollahi J, Carlsen IM, Mjaaland S, Skalle P, Rafiei A, Zarei S. Underbalanced drilling as a tool for optimized drilling and completion contingency in fractured carbonate reservoirs. SPE/IADC paper 91579 presented at the 2004 SPE/IADC Underbalanced Technology Conference and Exhibition held in Houston, Texas, USA; October 11−12, 2004.

[5] Prensky SE. Advances in borehole imaging technology and applications. In: Lovell MA, Williamson G, Harvey PK, editors. Borehole imaging: applications and case histories. London: Geological Society; 1999. p. 1−43.

[6] Buechler S, Ning J, Bharadwaj N, Kulkarni K, DeValve C, Elsborg C, et al. Root cause analysis of drilling lost returns in injectite reservoirs. SPE paper 174848 presented at the SPE Annual Technical Conference and Exhibition held in Houston, Texas, USA; September 28−30, 2015.

[7]   Adamson K, Birch G, Gao E, Hand S, Macdonald C, Mack D, et al. High-pressure, high-temperature well construction (Summer) Oilfield Rev 1998:36—49.

[8]   Dupriest FE. Fracture closure stress (FCS) and lost returns practices. SPE/IADC paper 92192 presented at the SPE/IADC Drilling Conference held in Amsterdam, The Netherlands; February 23—25, 2005.

[9]   Lavrov A, Larsen I, Bauer A. Numerical modeling of extended leak-off test with a pre-existing fracture (in print). Rock Mechanics and Rock Engineering 2015.

[10] Fjær E, Holt RM, Horsrud P, Raaen AM, Risnes R. Petroleum related rock mechanics. 2nd ed. Amsterdam: Elsevier; 2008.

[11] Lavrov A. The Kaiser effect in rocks: principles and stress estimation techniques. Int J Rock Mech Min Sci 2003;40(2):151—71.

[12] Raaen AM, Horsrud P, Kjørholt H, Økland D. Improved routine estimation of the minimum horizontal stress component from extended leak-off tests. Int J Rock Mech Min Sci 2006;43(1):37—48.

[13] Raaen AM, Skomedal E, Kjørholt H, Markestad P, Økland D. Stress determination from hydraulic fracturing tests: the system stiffness approach. Int J Rock Mech Min Sci 2001;38(4):529—41.

[14] Økland D, Gabrielsen GK, Gjerde J, Sinke K, Williams EL. The importance of extended leak-off test data for combatting lost circulation. SPE/ISRM paper 78219 presented at the SPE/ISRM Rock Mechanics Conference held in Irving, Texas; October 20—23, 2002.

[15] Caughron DE, Renfrow DK, Bruton JR, Ivan CD, Broussard PN, Bratton TR, et al. Unique crosslinking pill in tandem with fracture prediction model cures circulation losses in deepwater Gulf of Mexico. IADC/SPE paper 74518 presented at the IADC/SPE Drilling Conference held in Dallas, Texas; February 26—28, 2002.

[16] Power D, Ivan CD, Brooks SW. The top 10 lost circulation concerns in deepwater drilling. SPE paper 81133 presented at the SPE Latin American and Caribbean petroleum Engineering Conference held in Port-of-Spain, Trinidad, West Indies; April 27—30, 2003.

[17] Adachi J, Bailey L, Houwen OH, Meeten GH, Way PW, Schlemmer RP. Depleted zone drilling: reducing mud losses into fractures. IADC/SPE paper 87224 presented at the IADC/SPE Drilling Conference held in Dallas, Texas, USA; March 2—4, 2004.

[18] Rehm B, Schubert J, Haghshenas A, Paknejad AS, Hughes J, editors. Managed pressure drilling. Houston, Texas: Gulf Publishing Company; 2008.

[19] Liétard O, Unwin T, Guillot DJ, Hodder MH. Fracture width logging while drilling and drilling mud/loss-circulation-material selection guidelines in naturally fractured reservoirs. SPE Drill Completion 1999;14(3):168—77.

[20] Majidi R, Miska SZ, Yu M, Thompson LG, Zhang J. Quantitative analysis of mud losses in naturally fractured reservoirs: the effect of rheology. SPE Drill Completion 2010;25(4):509—17.

[21] Willson SM, Edwards S, Heppard PD, Li X, Coltrin G, Chester DK, et al. Wellbore stability challenges in the deep water, Gulf of Mexico: case history examples from the Pompano Field. SPE paper 84266 presented at the SPE Annual Technical Conference and Exhibition held in Denver, Colorado, USA; October 5—8, 2003.

[22] Gradishar J, Ugueto G, van Oort E. Setting free the bear: the challenges and lessons of the Ursa A-10 deepwater extended-reach well. SPE Drill Completion 2014;29(2):182—93.

[23] Fuh G-F, Morita N, Boyd PA, McGoffin SJ. A new approach to preventing lost circulation while drilling. SPE paper 24599 presented at the 67th Annual Technical Conference and Exhibition of the Society of Petroleum Engineers held in Washington, DC; October 4—7, 1992.

[24] Fuh G-F, Beardmore D, Morita N. Further development, field testing, and application of the wellbore strengthening technique for drilling operations. SPE/IADC paper 105809 presented at the 2007 SPE/IADC Drilling Conference held in Amsterdam, The Netherlands; February 20–22, 2007.

[25] Growcock FB, Alba A, Miller M, Asko A, White K. Drilling fluid maintenance during continuous wellbore strengthening treatment. AADE-10-DF-HO-44 paper presented at the 2010 AADE Fluids Conference and Exhibition held at the Hilton Houston North, Texas; April 6–7, 2010.

[26] Guo Q, Cook J, Way P, Ji L, Friedheim JE. A comprehensive experimental study on wellbore strengthening. IADC/SPE paper 167957 presented at the 2014 IADC/SPE Drilling Conference and Exhibition held in Fort Worth, Texas, USA; March 4–6, 2014.

[27] Aston MS, Alberty MW, McLean MR, de Jong HJ, Armagost K. Drilling fluids for wellbore strengthening. IADC/SPE paper 87130 presented at the IADC/SPE Drilling Conference held in Dallas, Texas, USA; March 2–4, 2004.

[28] Alberty MW, McLean MR. A physical model for stress cages. SPE paper 90493 presented at the SPE Annual Technical Conference and Exhibition held in Houston, Texas; September 26–29, 2004.

[29] Wang H, Sweatman R, Engelman B, Deeg W, Whitfill D, Soliman M, et al. Best practice in understanding and managing lost circulation challenges. SPE Drill Completion 2008;23(2):168–75.

[30] van Oort E, Razavi SO. Wellbore strengthening and casing smear: the common underlying mechanism. IADC/SPE paper 168041 presented at the 2014 IADC/SPE Drilling Conference and Exhibition held in Fort Worth, Texas, USA; March 4–6, 2014.

[31] Morita N, Black AD, Fuh G-F. Theory of lost circulation pressure. SPE paper 20409 presented at the 65th Annual Technical Conference and Exhibition of the Society of Petroleum Engineers held in New Orleans, LA; September 23–26, 1990.

[32] Morita N, Black AD, Fuh G-F. Borehole breakdown pressure with drilling fluids—I. Empirical results. Int J Rock Mech Min Sci Geomech Abstr 1996;33(1):39–51.

[33] Morita N, Fuh G-F, Black AD. Borehole breakdown pressure with drilling fluids—II. Semi-analytical solution to predict borehole breakdown pressure. Int J Rock Mech Min Sci & Geomech Abstr 1996;33(1):53–69.

[34] Kaageson-Loe N, Sanders MW, Growcock F, Taugbøl K, Horsrud P, Singelstad AV, et al. Particulate-based loss-prevention material—the secrets of fracture sealing revealed! SPE Drill Completion 2009;24(4):581–9.

[35] Aston MS, Alberty MW, Duncum S, Bruton JR, Friedheim JE, Sanders MW. A new treatment for wellbore strengthening in shale. SPE paper 110713 presented at the 2007 SPE Annual Technical Conference and Exhibition held in Anaheim, California, USA; November 11–14, 2007.

[36] Sanders MW, Young S, Friedheim JE. Development and testing of novel additives for improved wellbore stability and reduced losses. AADE-08-DF-HO-19 paper presented at the 2008 AADE Fluids Conference and Exhibition held at the Wyndam Greenspoint Hotel, Houston, Texas; April 8–9, 2008.

[37] Kumar A, Savari S. Lost circulation control and wellbore strengthening: looking beyond particle size distribution. AADE-11-NTCE-21 paper presented at the 2011 AADE National Technical Conference and Exhibition held at the Hilton Houston North Hotel, Houston, Texas; April 12–14, 2011.

[38] Finger J, Blankenship D. Handbook of best practices for geothermal drilling. Sandia National Laboratories; 2010. Contract No.: SAND2010–6048.

[39] Ghalambor A, Salehi S, Shahri MP, Karimi M. Integrated workflow for lost circulation prediction. SPE paper 168123 presented at the SPE International Symposium and Exhibition on Formation Damage Control held in Lafayette, Louisiana, USA; February 26−28, 2014.

[40] Tessari RM, Warren TM, Jo JY. Drilling with casing reduces cost and risk. SPE paper 101819 presented at the 2006 SPE Russian Oil and Gas Technical Conference and Exhibition held in Moscow, Russia; October 3−6, 2006.

[41] Strickler RD, Moore D, Solano P. Simultaneous dynamic killing and cementing of a live well. IADC/SPE paper 98440 presented at the IADC/SPE Drilling Conference held in Miami, Florida, USA; February 21−23, 2006.

[42] IADC-UBO-MPD-Glossary. UBO & MPD Glossary. IADC [cited 2014]. Available from: www.iadc.org/wp-content/uploads/UBO-MPD-Glossary-Dec11.pdf.

[43] Jenner JW, Elkins HL, Springett F, Lurie PG, Wellings JS. The continuous-circulation system: an advance in constant-pressure drilling. SPE Drill Completion 2005;20(3):168−78.

[44] Dow B, Baker J, Spriggs P, Voshall A. Anatomy of MPD failures: lessons learned from a series of difficult wells. SPE/IADC paper 163546 presented at the 2013 IADC/SPE Drilling Conference and Exhibition (DC) held in Amsterdam, NL; March 5−7, 2013.

[45] Terwogt JH, Mäkiaho LB, van Beelen N, Gedge BJ, Jenkins J. Pressured mud cap drilling from a semi-submersible drilling rig. SPE/IADC paper 92294 presented at the SPE/IADC Drilling Conference held in Amsterdam, The Netherlands; February 23−25, 2005.

[46] Fossli B, Sangesland S. Controlled mud-cap drilling for subsea applications: well-control challenges in deep wells. SPE Drill Completion 2006;21(2):133−40.

[47] Nakagawa EY, Santos H, Cunha JC. Application of aerated-fluid drilling in deepwater. SPE/IADC paper 52787 presented at the 1999 SPE/IADC Drilling Conference held in Amsterdam, Holland; March 9−11, 1999.

[48] Maranuk C, Rodriguez A, Trapasso J, Watson J. Unique system for underbalanced drilling using air in the Marcellus shale. SPE paper 171024 presented at the SPE Eastern Regional Meeting held in Charleston, WV, USA; October 21−23, 2014.

[49] Nelson EB, Guillot D. Well cementing. 2nd ed. Schlumberger; 2006. p. 773.

[50] Islam MA, Skalle P, Tantserev E. Underbalanced drilling in shale—perspective of factors influences mechanical borehole instability. IPTC paper 13826 presented at the International Petroleum Technology Conference held in Doha, Qatar; December 7−9, 2009.

[51] Al Maskary S, Abdul Halim A, Al Menhali S. Curing losses while drilling & cementing. SPE paper 171910 presented at the Abu Dhabi International petroleum Exhibition and Conference held in Abu Dhabi, UAE; November 10−13, 2014.

[52] Glynn-Morris T, King T, Winmill R. Drilling history and evolution at Wairakei. Geothermics 2009;38:30−9.

[53] Ramos SA. Aerated drilling. Presented at "Short course on geothermal development and geothermal Wells" organized by UNU-GTP and LaGeo, in Santa Tecla, El Salvador; March 11−17, 2012. p. 1−7.

[54] Sveinbjornsson BM, Thorhallsson S. Drilling performance, injectivity and profuctivity of geothermal wells. Geothermics 2014;50:76−84.

[55] Mokhalalati T, Reiley R. Aerated mud drilling experience in Abu Dhabi. SPE paper 36301 presented at the 7th ADIPEC, Abu Dhabi; October 13−16, 1996.

[56] Filippov A, Mack R, Cook L, York P, Ring L, McCoy T. Expandable tubular solutions. SPE paper 56500 presented at the 1999 SPE Annual Technical Conference and Exhibition held in Houston, Texas; October 3−6, 1999.

[57] Grant T, Bullock M. The evolution of solid expandable tubular technology: lessons learned over five years. OTC paper 17442 presented at the 2005 Offshore Technology Conference held in Houston, TX, USA; May2−5, 2005.

[58] Stringer JA, Farley DB. The evolution of expandables: a new era of monobore expandable well-construction systems. SPE paper 164171 presented at the SPE Middle East Oil and Gas Show and Conference held in Manama, Bahrain; March 10−13, 2013.

[59] Goncalves R, Tajalie AF, Meyer A. Overcoming lost circulation while cementing riserless tophole in deepwater. OTC-24764-MS paper presented at the offshore technology Conference Asia held in Kuala Lumpur, Malaysia; March 25−28, 2014.

[60] El-Hassan H, Abdelrahman M, Johnson C, Belmahi A, Rishmani L, Jarouj H. Using a combination of cement systems to defeat severe lost-circulation zones. SPE paper 92374 presented at the 14th SPE Middle East Oil & Gas Show and Conference held in Bahrain International Exhibition Centre, Bahrain; March 12−15, 2005.

[61] Isgenderov I, Bogaerts MVO, Kurawle I, Kanahuati AI, Aghaguluyev J. ECD management solves lost circulation issues in cementing narrow pressure window: a case study. IADC/SPE paper 170491 presented at the IADC/SPE Asia Pacific drilling technology Conference held in Bangkok, Thailand; August 25−27, 2014.

Chapter 7

Curing the Losses

Once losses have started, a common way to deal with the problem is to create a seal that will reduce the permeability of the thief zone. This will reduce mud losses so that drilling can continue or casing can be set. The seal can be created, for example, by injecting a settable material or a particle-fluid mixture into the formation.

When losses have occurred, knowing the location of the loss zone is important for placing the treatment. Losses often occur not at the drill bit but higher up the hole, or even at the very shoe of the last casing string set in the well. A previously sealed loss zone may reopen and cause losses [1]. Some techniques used for locating loss zones were discussed in Chapter "Mechanisms and Diagnostics of Lost Circulation."

The properties of the formation, such as the permeability and the fracture and pore sizes, affect the treatment outcome. This leads to inconsistent performance of treatments in different formations and is one of the reasons why no cure-all solution against lost circulation has been found yet.

Hundreds of treatments are available on the market. The goal of this chapter is not to present a complete list of available products, but to give a broad overview of how different types of treatments work and in what situations they could or could not be effective. A review and comparison of some commercially available products can be found, for example, in Ref. [2].

7.1 CLASSIFICATION OF TREATMENTS

Systems currently used in the industry to treat losses can be subdivided into the following groups (Fig. 7.1):

- settable systems;
- lost circulation materials, commonly known as LCMs (synthetic graphite, calcium carbonate, ground walnut shells, mica flakes, gilsonite, fibers, etc.);
- blends of settable materials and LCMs.

A variety of LCMs have been proposed and used in the industry over the past 100 years. These include chicken feathers, prairie hay, alfalfa pellets, ground battery casings, ground tires, cotton seed hulls, hog hair, etc. Examples of commercially available LCMs are calcium carbonate (ground marble), nut

Lost Circulation. http://dx.doi.org/10.1016/B978-0-12-803916-8.00007-8

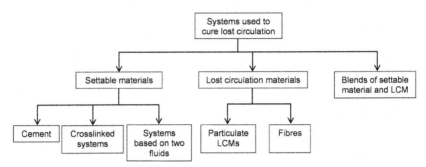

FIGURE 7.1   Classification of treatments.

hulls, graphite, petroleum coke, etc. LCMs can be added to the drilling fluid and, during well cementing, to the spacer or cement. They can also be applied as part of a lost circulation pill squeezed into the formation.

Settable materials work by building up a solid seal in the thief zone; eg, a cement seal. Slurries of settable materials usually are easily pumpable since they often contain little or no solids. Moreover, settable slurries can enter fractures and voids of any width, unlike LCM particles that can only enter fractures or voids of sufficient width for the particles to pass.

Since a given LCM can only bridge and seal fractures up to a certain maximum width, there is an effective operational range of fracture apertures for which each LCM is effective. Settable systems may enter narrower fractures and pores. The resulting seal is, however, often insoluble, which prevents the use of settable systems in pay zones. Moreover, premature gelation may reduce their sealing capacity.

Whatever treatment is used, it must be compatible with the fluids used in the well. It should also be able to pass through constrictions in the bottomhole assembly. In production intervals, the treatment should have as little impact on the formation permeability as possible.

It is believed that the effective use of granular particulates and flakes is limited to seepage losses or partial losses, and that they are considerably less effective at stopping severe and total losses. Crosslinked polymers, gunk, and cement have been used to combat severe losses. High-fluid-loss, high-strength particulates are reportedly effective at stopping severe losses, too [3].

Each of the solutions has thus strengths and weaknesses which make it work in one situation and fail in another. As a consequence, no universal LCM treatment that would be effective in all formations and operational conditions has been found yet. In particular, LCMs containing sized solids are usually ineffective against losses in cavernous (vugular) formations. Crosslinked systems might be effective in vugs, but might not develop sufficient compressive strength.

Since the primary objective of a treatment is to seal the high-permeability zones and fractures and thereby reduce the losses, it usually has a negative impact on the formation permeability. In some cases, permeability in the pay zone can be restored by an acid treatment after completion of the well; eg, in the case of an acid-soluble LCM. Acid treatments involve environmental and operational risks, require extra time, and do not always remove 100% of the damage (especially if the LCM is oil-wet) [4]. An alternative is a self-degradable LCM that remains in place for the time required to drill and complete the well, and decomposes thereafter.

The requirements imposed on lost circulation treatment systems are often mutually exclusive and can be summarized as follows:

- The material should be easily pumpable through the downhole equipment.
- The treatment should be compatible with the fluids in the well.
- The treatment should create an effective seal in the thief zone.
- The seal should be strong enough to withstand wellbore pressure fluctuations without being displaced inside the fractures or voids.
- The material placed against the productive interval should be removable by, eg, an acid treatment or by self-degradation before production starts.

Choosing the treatment is usually formalized in form of a decision tree. Examples of such decision trees can be found in refs. [5,6]. Decision trees used by operators are based on the available knowledge and understanding of lost circulation mechanisms as well as on the unique experiences of specific operators. Since decision trees are operator-specific, no universally accepted and standardized such tree has been developed so far. Therefore, no decision tree is provided in this book. Instead, strengths and weaknesses of different treatments are discussed in this chapter, and suggestions for treating lost circulation are provided, based on the mechanisms and severity of losses.

7.2 HOW DOES LOST CIRCULATION MATERIAL WORK?

LCM is a particulate or fibrous material pumped downhole in order to bridge and seal fractures and voids, and thereby stop losses. Granular materials (particulates, including flakes) and fibers have been used as LCM. Blends of fibers with particulates are common as well. Particulate LCMs include granular and flaky materials (Some authors divide LCMs into granular, flaky, and fibers rather than into particulates and fibers.). There is currently no universally accepted classification of LCMs, although some efforts to introduce such a classification have been made; see, eg, Ref. [2].

The seal formed by LCM can build up either deep inside the fracture or near its mouth, ie, near the borehole wall. If LCM particles are deposited on the borehole wall without entering the fracture, the seal can be easily destroyed by the drillstring action. Such an unstable seal can be unintentionally created if LCM particles are too large to enter the fracture. Therefore, LCM particles must be small enough to enter the fracture but large enough to build the bridge.

In addition, the material must be pumpable; ie, the particles should not block the conduits on their journey downhole. The design of LCM is therefore an optimization exercise where fracture characterization, especially fracture width (aperture), provides valuable input data.

The lifecycle of an LCM is sometimes divided into the following stages [7]:

- dispersion;
- bridging;
- sealing;
- sustaining.

An effective application of LCM depends on its performance during each of these stages.

Dispersion of LCM in fractures or vugs means that the LCM should, firstly, be delivered to the place where it is intended to be deposited, without getting stuck or screened out along the way and without plugging the surface pumps or the bottomhole equipment.

Bridging means that a mechanical bridge is created across the fracture on which a low-permeability seal can then be built. The bridge provides the mechanical strength needed to sustain the pressure gradient across the subsequently formed seal.

Sealing means creating a low-permeability flow barrier (*seal*) by depositing fine particles on the bridge. Fine particles effectively form a filter cake on the bridge as the mud is defluidized.

Sustaining mechanical and hydraulic loads (erosion, pressure gradient) is essential for the ability of the seal to perform its functions for the time sufficient to drill through the lost circulation zone and cement the well. Therefore, strength of the seal is important. This is usually characterized by the *shear strength* or *unconfined compressive strength (UCS)*.

The distinction between a bridge and a seal is as follows: The bridge provides mechanical strength, but its permeability is too high since it is formed by relatively coarse LCM particles. In order to stop the losses, the space between the particles in the bridge must be filled with small particles (either relatively fine LCM particles, or clay particles normally present in the drilling fluid). Filling the voids in the bridge by fine particles seals the conduit. Bridging and sealing are schematically illustrated in Fig. 7.2.

The main requirements to an effective LCM can be summarized as follows:

- LCM should be easy and fast to mix.
- LCM should be easy and safe to deploy.
- LCM should not block constrictions in the bottomhole assembly.
- LCM should defluidize rapidly once placed in the permeable formation.
- The seal created by LCM after defluidization should have sufficient mechanical strength to survive hydraulic gradients and mechanical loads during the time it is intended to.

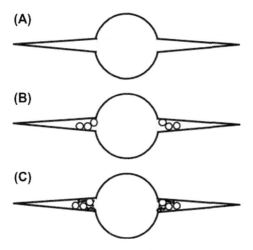

FIGURE 7.2 LCM action: bridging and sealing. (A) Fracture before application of LCM. (B) Coarse LCM particles bridge the fracture. (C) Fine particles (LCM, barite) are deposited on the bridge and seal the fracture.

- LCM should have sufficient thermal stability to sustain elevated downhole temperatures.
- All in all, LCM should be effective at fixing the lost circulation problem.

When added to cement to stop losses during well cementing, LCM additionally must be inert to the cement constituents. Particulate LCMs (eg, nut shells) and fibers (nylon, polypropylene) have been used with well cements [5].

Loss prevention materials (LPMs) used in wellbore strengthening and introduced in Chapter "Preventing Lost Circulation" are somewhat similar to LCM in action. Indeed, both LCM and LPM must create a seal in the fracture. However, using LCM for curing losses usually requires higher concentrations than using the same material as LPM to prevent losses [8].

If the main mechanism of wellbore strengthening is the fracture propagation resistance, then there is indeed very little difference between LCM and LPM. However, if the stress cage mechanism is at play in wellbore strengthening, the function of LPM becomes different from LCM. In particular, LPM must then be able to prop the fracture open and sustain compressive hoop stress when the wellbore pressure is reduced and the fracture tries to close. Many materials traditionally used as LCM are also used as LPM. The performance of both materials is quantified by their fluid loss and the maximum pressure difference that the LCM/LPM seal can sustain [9].

The type of the base fluid has a profound effect on the ability of LCM to cure losses. In particular, similar to LPMs (chapter: Preventing Lost Circulation), LCMs used in water-base muds (WBMs) have a better ability to build the seal than LCMs used in oil-base mud (OBM) or synthetic-base mud

(SBM) [10]. On the other hand, it is precisely the ability of WBM to build a better seal that increases the risk of differential sticking with such muds. This is another example of a tradeoff in designing lost-circulation treatments.

The type of the base fluid has also a profound effect on the shear strength of the LCM seal. Laboratory experiments demonstrate that the shear strength developed with an OBM is typically several times lower than with a WBM. Weighting an LCM-laden fluid with barite has a detrimental effect on the shear strength of the seal [3,11].

LCMs can be introduced into a normally circulated drilling fluid. LCMs can also be applied as a part of a squeeze treatment, eg, a high-fluid-loss squeeze. A high-fluid-loss squeeze system is a mixture of, eg, diatomaceous earth, a bridging agent (LCM), and barite. The base fluid can be either water or oil. The system can be pumped through the drillbit [5]. High-fluid-loss systems can be effective at curing severe losses in fractured or high-permeability rocks [2].

7.3 PARTICULATE LOST CIRCULATION MATERIALS

Particulate LCMs are typically mixed at the rig, pumped downhole, and placed in the thief zone to block the flow pathways into the formation. Effective bridging and sealing require that, once placed in the formation, the material be defluidized rapidly by fluid filtration into the rock [11].

Sized solids commonly used as LCM are ground walnut hulls, calcium carbonate (ground marble), synthetic graphite, gilsonite, perlite, asphalt, flake-type materials (eg, mica or cotton seed hulls), and others. The particle size of LCMs is typically around 250−600 μm. LCMs based on nanoparticles (1−100 nm in size) made of silica, iron hydroxide, barium sulfate, or calcium carbonate have been developed [2]. There seems to be a general consensus, supported through laboratory tests and field experience, that particulate LCMs with broad particle size distribution (PSD) perform better than monosized particles [9,12]. As mentioned earlier, fine particles fill the space between the larger ones, creating a tighter seal. But there must be sufficient amount of large particles to bridge the fracture, in the first place.

A broad PSD also enables bridging fractures of different apertures. This is especially important when dealing with losses caused by natural fractures. An example of LCM with several groups of particle size ranges was described by Savari and Whitfill [13]. Such multimodal particle systems can be used either on their own or in combination with fibers or swellable polymers.

Other factors affecting the performance of particulate LCMs include the type of the base fluid, the shape of the particles, the size of the openings to be sealed (fractures, vugs) relative to the particle size, and the concentration of LCM. In particular, flake-shaped particles perform worse than calcium carbonate particles, at least in laboratory tests [14]. Flake-shaped particles have relatively low mechanical strength and therefore are more easily crushed during transport and placement [15]. Irregularly shaped, deformable particles, such as nut shells, perform quite well at sealing even wide fractures (up to 5 mm aperture) [9].

Properly designed blends of several LCMs can be more effective than each material on its own. An example of such an effective system is a blend of resilient graphite, sized calcium carbonate, and fibers, in the proportion 1 to 2 to 0.25 by weight, with a total concentration of 40−60 lb/bbl (114−171 kg/m^3) [8].

Higher LCM concentration is usually required to cure more severe losses and to seal wider fractures. However, increasing the LCM concentration may eventually jeopardize the ability of LCM to penetrate and bridge the fractures [9]. Example concentrations of particulate LCMs in drilling fluids are 10−40 lb/bbl (28.5−114 kg/m^3) [5].

Particles used as an LCM are known to decrease in size and to change shape with time, as they are circulated. This is caused by particle impact on the bottom of the hole, particle collisions with the impeller of the centrifugal pump and with each other, etc. Laboratory tests have been used to estimate the degree of degradation of different particle types. For instance, in the tests described in Ref. [16], the drilling fluid carrying particles was sheared in a mixer at 7000 rpm for some time (from 5 to 15 min). The weight of particles of specific fractions was measured before and after the test. The shear degradation tests have revealed that the degree of degradation of granular materials increases with the shear rate and with the time of shearing. Also, the particle shape changes, particles becoming more rounded. The size degradation is more severe for larger particles. Out of three materials (ground marble, graphite, walnut hulls), ground marble is worst in terms of degradation. Adding large quantities of barite into the fluid somewhat protects ground marble from degradation. Barite itself degrades even more than ground marble but does not affect the degradation of walnut hulls [16]. Based on the laboratory tests, calcium carbonate is not a very suitable material for circulation as part of the drilling mud. Graphite and walnut hulls, on the other hand, survive shearing quite well [16].

Even though LCMs in many cases are effective at curing lost circulation, excessive particle load in the drilling fluid may increase the risk of differential sticking and cause poor hole cleaning [14]. The latter is one of the factors that may increase the bottomhole pressure (BHP), making losses worse.

Another problem with LCMs is their inconsistent performance, even in laboratory experiments. For instance, some graphitic LCMs increase the fracture reopening pressure, while others decrease it, according to the laboratory results reported by Adachi et al. [17]. The authors attributed these differences to the different resiliency of LCMs. This issue, however, needs further research. Inconsistent performance of LCMs is also known from field applications.

7.4 FIBERS

The main idea of using fibers as LCM is to create a network structure, a mat-like bridge that would facilitate deposition of finer particles. Different materials have been used as fibers; eg, cellulose, nylon, polypropylene, shredded paper, etc. Acid-soluble fibers have been developed for use in pay zones.

Concentration of fibers in the slurry should be sufficiently high to create such a bridge. Making the concentration too high may, however, jeopardize the pumpability of the fluid since it increases its viscosity [8]. Being able to pump the fiber-laden LCM through the surface and bottomhole equipment is crucial for its successful deployment.

Fibers are usually used in combination with particulate materials such as calcium carbonate (ground marble) or graphite. The basic principle is that fibers create a bridge onto which the granular material can be deposited. The intended result is a low-permeability seal. The PSD of the granular material deposited on the bridge should match the size distribution of the voids in the structure formed by the fibers [7].

An 50-bbl pill used to treat partial losses could have the following composition [18]:

- fine calcium carbonate (10 lb/bbl; ie, 28.5 kg/m^3);
- coarse calcium carbonate (10 lb/bbl; ie, 28.5 kg/m^3);
- fine or medium fibers (10 lb/bbl; ie, 28.5 kg/m^3).

The length of an individual fiber can be, for example, 1500 to 2000 μm (viscose cellulosic fibers), 3000 μm (carbon fibers), or 180 μm (oil-coated cellulosic fibers) [10]. The viscosity of the slurry increases with the fiber length. The aspect ratio of the fiber,—ie, the length-to-diameter ratio—varies widely and can be equal to, eg, 30 to 35 (viscose cellulosic fibers), 50 (carbon fibers), or 1.8 (oil-coated cellulosic fibers) [10].

In laboratory experiments with a WBM, a combination of ground marble, resilient graphitic carbon, and fibers successfully sealed the fracture and significantly reduced mud losses [10]. In the same experiments, blends of ground marble with resilient graphitic carbon, or ground marble with fibers, could not bring the losses under control. Moreover, the performance of fibers was significantly poorer with a nonaqueous base fluid.

Based on laboratory tests, the optimum concentration of viscose fibers and oil-coated fibers in the water-base fluid was found to be 1.8 lb/bbl (5.13 kg/m^3) and 4 lb/bbl (11.4 kg/m^3), respectively [10]. Hence, the optimum concentration of fibers decreases with their length. In addition to the length and aspect ratio, the performance of fibers depends on their mechanical properties (stiffness), concentration, fluid type and viscosity, and the fracture morphology (roughness and aperture) [4,7].

The strength of the bridge formed by fibers is critical for the subsequent ability of the seal to sustain a pressure gradient and remain in place. In order to increase the strength of the bridge, a dual-fiber formulation has been proposed for use with WBMs [7]. The dual-fiber material is a blend of relatively soft fibers and relatively rigid fibers. The strength of the resulting bridge arguably increases because of the contrast in stiffness. The dual-fiber system discussed in [7] can reportedly seal fractures up to 5 mm wide, and the resulting seal can sustain a pressure difference of 2000 psi (13.8 MPa). The material was successfully used in the tophole section of a deepwater well as part of a spacer formulation before cementing the surface casing in a lost circulation zone.

As with other LCMs, fibers used to treat mud losses in pay zones have a negative impact on future productivity. A solution, in the case of water-base fluids, is to use self-degradable fibers. Such fibers stay in place during drilling and completion of the well, and degrade by the time production starts. The degradation time can be controlled by varying the pH of the fluid. Fibers described in [4] are applicable within the temperature range of 40°C−85°C, with stability time ranging from a few days to a few weeks. The volumes of LCM pills containing degradable fibers were 4−8 m^3 in the field cases quoted in [4].

Mechanical properties of the seal formed by a fiber-laden LCM may be quite different from a pure particulate LCM. In particular, laboratory tests revealed substantial compressibility or ductility of fiber systems. Sometimes, it is even difficult to identify the UCS of these materials [11].

Fibers are often used as a lost-circulation additive in well cement. In this application, fibers also improve the cement strength by inhibiting crack propagation in hardened cement.

7.5 HOW DOES SETTABLE MATERIAL WORK?

A settable material is a material that is pumped in liquid state and solidifies downhole, sealing the thief zone. Examples of settable materials are cement, gunk, bentonite−oil−mud systems, cross-linked systems, etc. Before setting, settable materials should have sufficiently low apparent viscosity in order to be easily pumpable. However, effective displacement of drilling fluid in fractures requires that the settable material should be more viscous than the drilling fluid in order to fully displace and replace the latter in the fractures and high-permeability thief zones.

Upon setting, the material should develop sufficient strength to be able to withstand pressure variations in the wellbore. Sealing capacity of settable materials can be increased if they exhibit swelling behavior. Swelling results in tighter sealing of fractures and may increase the strength of the seal.

An important property of the set material is its yield stress. The yield stress determines the mechanical stability of the seal in a fracture when a pressure differential is applied across the seal. The maximum pressure gradient that the seal can sustain during, eg, a pressure surge in the well is given by [19]:

$$|\nabla P|_{max} = \frac{2\tau_Y}{w_h} \qquad [7.1]$$

where w_h is the hydraulic aperture of the fracture; τ_Y is the yield stress of the sealing material. The ability of the settable material to sustain pressure surges is thus determined by its yield stress upon setting, according to Eq. [7.1].

7.6 CEMENT

Sealing with cement can be effective in those zones where other treatments would fail. Cement's compressive strength is higher than that of other settable systems—eg, gunk—which improves the quality of the seal. Cement was recommended as a cure against severe losses in permeable sandstones [14]. Cement is one of the few treatments that are reportedly effective against losses in vugular formations [20,21]. Bentonite, particulate LCMs (eg, calcium carbonate), and fibers can all be part of cement systems used to cure losses.

If the well is drilled with nonaqueous drilling fluid, a sequence of spacers needs to be pumped before cement, in order to water-wet the formation surfaces exposed in the annulus and to prevent cement from mixing with the drilling fluid (Box 7.1).

BOX 7.1 Curing Losses With Cement

The following example of using cement to stop losses in fractured and vuggy carbonates was provided by Fidan et al. in their IADC/SPE paper 88805 "Use of cement as lost circulation material—field case studies."

The well was drilled with a potassium formate OBM (mud density 0.97–1.05 g/cm³). A thief zone was located at 1000 to 1300 m true vertical depth. The decision was made to plug the thief zone with thixotropic cement. The following fluid train was pumped:

1. 1.5 m³ of diesel spacer with a surfactant (6 L/m³) and a mutual solver (6 L/m³);
2. 1.5 m³ of tuned spacer (density 1100 kg/m³);
3. 9 m³ of low-density thixotropic cement containing flaky LCM.

The two spacers pumped first served to prevent cement from mixing with the OBM present in the annulus. They also provided water-wetting, which was essential to ensure high sealing quality.

Cement was spotted against the thief zone in the well, and a squeeze was performed. After setting, the cement plug was drilled through, and drilling continued without losses.

The same procedure was successfully applied in seven other wells in the same field.

Particularly suitable for cement squeeze treatments are thixotropic cements [22]. These cements flow easily, but rapidly develop gel strength when the pumping stops. The gel strength prevents the cement from falling back. Thixotropic properties also reduce the gas migration. Rapid buildup of gel strength prevents cement from invading the formation too deeply, thereby reducing the formation damage. Thixotropic cements are effective against severe losses in naturally fractured rocks [5].

Crosslinked cement is defined as "a combination of cement and frac products mixed as regular cement slurry in a gelled fluid" [23]. Two versions of such cement are described in Ref. [23]: acid-soluble magnesia cross-linked cement and a regular cross-linked cement. Due to its solubility, the magnesia cross-linked cement is preferred over the regular cross-linked cement in pay zones. The regular cross-linked cement is a cost-effective alternative to the magnesia cross-linked cement and may be used in nonproductive intervals.

In one of the field applications described in [23], the magnesia cross-linked cement was used to seal a fractured zone and enable the operator to do a cementing job without losses. The amount of magnesia cross-linked cement squeezed into the formation was on the order of 0.5—1.0 L per meter well.

The lost circulation capacity of cement can be impaired by channels created in cement by mud or formation fluids before the setting is complete. Another disadvantage of cement is irreversible formation damage. Moreover, placing a cement plug and waiting for it to set may increase the nonproductive time significantly, up to several days [13].

7.7 SETTABLE PILLS BASED ON CROSS-LINKED SYSTEMS

Crosslinking is the linking of two polymer chains by a cross-linking agent. The cross-linking agent is activated by time and temperature or by shearing at the drill bit. After setting, the pill produces a rubbery, ductile substance sealing the fracture and preventing further losses. The setting time is typically a few hours [24]. The setting time can be controlled by adding retarders or accelerators, depending on the formation temperature. Retarders are used when a longer pumping or setting time is required, and in formations with elevated temperatures. The use of retarders may be required in order to prevent setting inside the drill string while treating thief zones located at great depth [20]. Accelerators are used to reduce the setting time and in formations with lower temperatures.

A typical cross-linked pill is a blend of polymers, cross-linking agents, and LCMs. A squeeze with a cross-linked system can be effective against losses in depleted sands and unconsolidated formations [20].

A blend of a cross-linked polymer and fibrous cellulose was used in deepwater wells in the Gulf of Mexico drilled with an SBM (Box 7.2) [24]. This pill could reportedly be used with any water-, oil- or synthetic-based mud. The use of the pill in pay zones may irreversibly reduce the permeability since

BOX 7.2 Application of a Cross-linked System in the Gulf of Mexico

The following examples of using cross-linked pills in deepwater Gulf of Mexico wells were described by Caughron et al. in their IADC/SPE paper 74518 "Unique crosslinking pill in tandem with fracture prediction model cures circulation losses in deepwater Gulf of Mexico."

In one well, the formation fractured below the 13-5/8" casing set at 17,703 ft (5396 m). The type of losses (induced fractures) was determined from the equivalent circulating density (ECD) buildup signatures (rapid ECD buildup before fracturing, but gradual buildup after fracturing). An attempt to cure the losses by adding LCM to the drilling fluid failed. Therefore, a cross-linking pill was chosen. The first attempt to squeeze the pill, using cement pumps, failed. Rig pumps were then used to perform the squeeze. The flow rate during the squeeze was 1–3 bbl/min. Circulation was completely regained after the pill had set.

In another deepwater Gulf of Mexico well, two thief zones in the same open hole interval were identified, at 15,150 to 15,265 ft (4618 to 4653 m) and 15,750 to 15,809 ft (4801 to 4819 m). The lower thief zone was due to a fault. The upper thief zone was due to fractured shale. Sealing both zones simultaneously was not possible since it would have made the pumping and tripping time too long. Therefore, two cross-linking pills were squeezed, 110 bbl (17.5 m^3) each. Resistivity logs performed after the squeeze confirmed high quality of the seals. Drilling then continued without losses.

Among the lessons learned in these wells, Caughron et al. emphasize the importance of gradual pump rate decrease at the end of the squeeze, and gradual pump rate increase when drilling is resumed.

the pill is not degradable, thermally or chemically. According to Ref. [24], retarders should be used with this pill when the bottomhole temperature is above 38°C (100°F); accelerators should be used when ambient, water, or formation temperature is below 16°C (60°F). It is recommended to increase the flow rate gradually at the beginning of the squeeze, and to decrease it gradually at the end of it. This will reduce the pressure surge (at the beginning of the squeeze) and the amount of the fluid flowing back from the fracture (at the end) [24].

A blend of a cross-linking polymer and a fibrous material has been advocated as a remedy against total losses occurring in subsalt shales [25]. Particle distribution in such a pill can be adjusted so as to match the apertures of fractures. Upon setting, the pill makes a gel-like structure sealing the fractures. The pill can be mixed in advance; ie, before the lost circulation incident. Once lost circulation occurs, a cross-linking agent is added, and the pill is squeezed into the thief zone. A pumping rate of no more than 320 L/min was recommended in Ref. [25], with the volume of the pill 8 m^3. According to Ref. [25], the pill can be mixed in freshwater, saltwater, or seawater and can be used with WBM, OBM, or SBM.

The use of cross-linking pills is usually limited to nonproducing intervals since they are not degradable and not acid soluble. Moreover, performance of cross-linked systems depends on the downhole temperature. Uncertainty in the latter increases the uncertainty in the performance of these materials.

An example of a cross-linked system that can be used in reservoir sections was described in Ref. [19]. The material, called "nanocomposite organic/inorganic gel," contained four constituents:

- a polymer base;
- a cross-linking agent;
- swelling cross-linked polymer grains;
- colloidal particles (clay).

The yield stress of the set material was 3–5 kPa. Crosslinked polymer grains, swelling in contact with water, improve the quality of the seal. The material is soluble with acid or hydrogen peroxide.

7.8 SETTABLE PILLS BASED ON TWO FLUIDS

Two-fluid pills consist of two fluids, fluid No.1 and fluid No. 2, pumped separately. The two fluids set and build a seal after making contact with each other downhole. Up until the fluids reach the thief zone, they are kept separated from each other. To this end, a spacer can be used, or the two fluids can be pumped through different pathways (one through the drillstring, the other one through the annulus) [5]. The downhole mixing and activation require controlled flow rates. Therefore, such treatments cannot be used to cure total losses.

In the case of a well drilled with WBM, a bentonite—oil system can be used as fluid No.1. It consists of oil (diesel, mineral, or synthetic), bentonite, and, possibly, cement [5]. The bentonite—oil system is a settable material: it forms a plastic plug when it downhole comes in contact with fluid No. 2, which is water or WBM [19]. During the treatment, the drillpipe is raised just above the thief zone. Fluid No.1, the bentonite—oil slurry, is pumped down the open-ended drillpipe, while fluid No. 2, the WBM, is pumped down the annulus [1,26]. When mixed downhole, the slurry sets and can then be squeezed into the thief zone. It is imperative that the arrival of the two fluids (the bentonite—oil slurry and the mud) at the bottomhole is timed correctly. The properties of the final plug depend on the ratio of the drilling mud to the bentonite—oil slurry. Increasing this ratio results in a softer plug that is easier to squeeze into the formation. Reducing the ratio results in a firmer plug [18]. Optionally, LCM may be added to the bentonite—oil mixture. When preparing the slurry, it is crucial to avoid contamination with water or mud since this can cause premature gelation.

A disadvantage of this type of treatment is that the pill is difficult to spot precisely. Moreover, the treatment may violate environmental policies.

In addition, the seal creates significant formation damage, just like most other settable systems. Curing losses with bentonite–oil–WBM systems is extensively covered in Ref. [1].

An example composition of the bentonite–oil pill is given in Ref. [18]:

- bentonite: 300–400 lb (136–181 kg);
- diesel oil: 1 bbl (0.16 m^3);
- optional: LCM at 10–12 lb/bbl (28.5–34.2 kg/m^3);
- barite.

Examples of pump rates are as follows: bentonite–oil slurry at 1 to 2 bbl/min down the drillpipe, mud at 4 to 8 bbl/min down the annulus. The mud pump rate can be increased (eg, to 8–16 bbl/min) in order to obtain a softer plug [18].

When the well is drilled with OBM, a *reverse squeeze* is to be used. In this case, the pill is made of organophillic bentonite mixed with water. Similarly to the application of the bentonite–oil system, the drillpipe is raised just above the thief zone, and the organophillic–bentonite–water slurry is pumped down the open-ended drillpipe, while mud (OBM, in this case) is pumped down the annulus. When the slurry comes in contact with the OBM downhole, a plug is produced that can then be squeezed into the formation to repair the thief zone [18].

7.9 CURING LOSSES CAUSED BY HIGH-PERMEABILITY MATRIX

Sufficiently large pores in the rock matrix can create enough permeability to cause losses (10…100 D). In this case, solids normally present in the drilling fluid (barite and bentonite particles) will not be able to seal the pores. In particular, if solids in the drilling mud are smaller than c. 1/3 of the pore size, they will be carried by the fluid into the pore space. Hence, they will not be able to create a filter cake on the borehole wall, and losses will continue. Depending on the permeability of the rock and properties of the drilling fluid, the severity of such losses can vary from seepage to total losses [5]. This demonstrates that the severity of losses is a poor indicator of lost circulation mechanisms.

Examples of high-permeability formations that may cause lost circulation are unconsolidated sands and gravels at shallow depth. Lost circulation is such formations may lead to washouts, jeopardizing the wellbore stability and the quality of subsequent cementing.

The PSD of LCM used to treat losses caused by high-permeability matrix must be such that the pore throats are plugged effectively. If LCM particles are too small, they will not be able to bridge the pore throats and will be carried into the porous media. If LCM particles are too large, they will be deposited on the borehole wall where they can be dislodged by the drillstring action, reexposing the formation to the drilling fluid. According to some estimates, 90% of particles must be smaller than the typical pore size in order for LCM to work in high-permeability porous zones [27]. Another rule of thumb stipulates that the median size of LCM particles, D_{50}, must be equal to or slightly greater than the median pore size. This gives rise to an approximate relationship between the rock permeability and the recommended median size of LCM. For instance, rock permeability of 520 mD would require an LCM with $D_{50} = 7.6$ μm, while permeability of 16.6 D would require $D_{50} = 43$ μm [5]. This results in a convenient, simple technique for choosing LCM based on D_{50} only. It does not, however, take into account the whole PSD and how it might affect different stages of LCM action; ie, bridging by coarse particles and sealing by fine particles. More advanced models—eg, those based on the ideal packing theory—can be invoked to tailor the LCM's PSD for a particular rock [5]. Some modern LCM designs put less emphasis on D_{50} and rather aim at optimizing the entire PSD. An example is the multimodal LCM described in Ref. [13]. D_{50} alone is not sufficient to fully characterize the sealing capacity of such materials.

Calcium carbonate and synthetic graphite particles have been used to combat seepage losses in porous matrix. They are used with both water- and synthetic-base drilling fluids [27]. Fiber-based LCM is another material reportedly effective against losses in high-permeability rocks [5,27].

7.10 CURING LOSSES IN VUGULAR FORMATIONS

Losses in vugular formations are among the most difficult lost circulation problems. Drilling blind, with no returns, is a possible strategy when drilling in vugular formations [5,28]. This strategy requires, however, that sufficient quantities of mud are available on the surface. Special drilling methods, such as underbalanced drilling, casing while drilling, and pressurized mud-cap drilling, can be used as well, as discussed in Chapter "Preventing Lost Circulation."

Curing losses with LCM in vugular formations is not an easy task. Introducing fibers in LCM pills has been suggested as a possible cure against lost circulation in such formations [10]. Another option is to use a cross-linking pill that forms a gel structure and can be effective in vugular rocks. The resulting seal is, however, not acid-soluble which prevents the use of such pills in productive zones.

7.11 CURING LOSSES CAUSED BY NATURAL FRACTURES

Mud losses caused by natural fractures are most common in carbonate formations, gas-bearing shales, and geothermal wells. Losses into wide natural fractures are among the most difficult to treat. Interconnected fracture networks often found in sedimentary rocks provide high-permeability escape routes for the drilling fluid. The challenge of designing treatments against losses in naturally fractured rocks is aggravated by unknown apertures of natural fractures. This is unlike induced fractures where decades of research in hydraulic fracturing have made it possible to provide at least rough estimates of the fracture aperture. In naturally fractured rocks, the hydraulic apertures depend on the fracture orientation, the magnitudes and orientation of in situ stresses, and fracture mineralization. Most of these variables are unknown. The fracture aperture plays an important role in designing the optimal PSD of LCM.

The use of LCM in naturally fractured reservoirs is a debated issue. On one hand, in many cases, LCM can stop losses. On the other hand, LCM damages the fracture permeability and is difficult to remove by acid treatment. Since fractures play crucial role in productivity of naturally fractured reservoirs, minimizing formation damage should be a priority in the pay zone. As pointed out by Liétard et al. [29], in naturally fractured reservoirs, *preventing* lost circulation by properly designing the drilling fluid rheology is advantageous to *curing* losses with LCM. Liétard et al. recommended that LCM should be "avoided as much as possible" in such rocks. Instead, effort should be made to prevent losses by increasing the yield stress of the drilling fluid and reducing the overbalance. If preventing lost circulation is successful, flowing back the mud bank left in the fracture is much easier than removing the LCM and is therefore advantageous from reservoir engineering point of view.

If preventive measures do not work, and the decision is made to use LCM, the particles should be sized in such way as to cure the losses but, at the same time, keep the formation damage to a minimum. The general idea of treating losses in naturally fractured rocks with LCM is that solid particles must have the optimal size (that size being a function of the fracture aperture) and should be deposited as close to the wellbore as possible. Sealing the fracture close to the wellbore will minimize the total volume of losses and will reduce the duration of the mud loss incident. The size of the particles should be such that they are sufficiently small to be able to enter the fracture (otherwise particles will only build a seal at the borehole wall which can subsequently be destroyed). Particles should, however, be sufficiently large to be deposited close to the wellbore rather than being transported deep into the fracture. As a rule of thumb, particles approximately 2.5 times smaller than the fracture aperture are often believed to be the optimal size for bridging natural fractures. In addition, the PSD should be such that the permeability of the deposited LCM is as low as possible. This can be achieved with a PSD in which smaller particles fill the space between large ones. In addition, a broad PSD ensures that fractures of different apertures can be sealed. Fractures of different apertures may be present in the same thief zone, or even the same fracture can have different apertures in different parts of its mouth (chapter: Mechanisms and Diagnostics of Lost Circulation). A broad PSD improves the chances that particles of the "right" size will be present in the LCM.

The issue of properly sizing LCM particles is aggravated by LCM wear as it travels downhole. Because of particle wear (grinding), the PSD introduced at the rig is different from the PSD that eventually reaches the fracture. Designing LCM resistant to wear is an important subject within the lost-circulation research and development.

Fibers are usually ineffective against losses caused by large-aperture fractures. Fibers in combination with particulates are reportedly effective at stopping losses into wide natural fractures [4,10]. It has been hypothesized that fibers are deposited at the entrance into the fracture, providing a structure on which other LCM particles can then be deposited. Fibers (along with other particles) should, however, be transported far enough into the fracture and be deposited inside the fracture. This will reduce erosion of the particle bridge by the action of the drillstring [4].

Particles of synthetic graphite are reportedly effective at sealing fractures. The effectiveness of synthetic graphite has been attributed to its ability to deform under compression. In addition, graphite is inert and does not react with drilling fluids [27]. Some experimental evidence suggests, however, that graphite might be inefficient as an LCM since it creates a lubrication effect in the bridge [9]. As such, adding graphite might indeed reduce the sealing capacity of the LCM system.

Another material that could reportedly seal fracture apertures of up to 3 mm are coarse cellulose flakes, up to 1.3 mm in size [30].

The use of pills based on cross-linking polymers, in combination with fibers, is reportedly effective in heavily fractured thief zones, such as subsalt shales [25]. The use of these materials in productive zones is limited since they create a seal that is neither degradable nor acid soluble.

7.12 CURING LOSSES CAUSED BY INDUCED FRACTURES

Fractures induced by drilling show less variety in their properties (eg, length and aperture) than preexisting, natural fractures do. Indeed, natural fractures could be created at any time in the geological history. Their properties are thus determined by the conditions that existed at that moment and by subsequent geological history. Induced fractures, on the other hand, are created during drilling, and many variables affecting their development (eg, the BHP) are known. In theory, it should then be possible to estimate the apertures of induced fractures and use this information to design the LCM. In practice, this requires (a) that the rock properties and in situ stresses are known and (b) that an accurate model of induced fracturing is available. Generally, neither of the two is the case.

Two classical models of induced fracturing are often used to estimate the parameters of drilling-induced fractures: the Perkins–Kern–Nordgren (PKN) model and the Khristianovitch–Geertsma–de Klerk (KGD) model.

Consider a vertical well in an impermeable formation. Assume that the minimum in situ stress is horizontal. Both models assume plane-strain conditions. In the PKN model, the plane-strain conditions are in the vertical plane normal to the direction of the fracture propagation. The model thus describes a long fracture of finite height (Fig. 7.3). In the KGD model, the plane strain is in the horizontal plane. The model thus represents an infinitely high fracture of finite length (Fig. 7.4). Both models have their origin in the theory of hydraulic fracturing [31]. The maximum fracture aperture is at the wellbore wall in both models.

In the PKN model, the fracture aperture at the borehole wall varies between zero (at the top and bottom of the fracture) to w_{max} (at midheight) (Fig. 7.3). The maximum fracture aperture is given by [31].

$$w_{max} = \frac{2H(P_w - \sigma_h)}{E'} \qquad [7.2]$$

where H is the fracture height; P_w is the wellbore pressure; σ_h is the minimum horizontal in situ stress; E' is the plane-strain modulus given by:

$$E' = \frac{E}{1 - v^2} \qquad [7.3]$$

where E and v are the Young's modulus and the Poisson's ratio of the rock.

The PKN model demonstrates that the fracture aperture at the wellbore does not need to be constant. This suggests the use of a broad PSD to combat mud losses. PKN geometry was employed by Adachi et al. in their numerical model of mud losses caused by *natural* fractures [17]. The use of PKN

FIGURE 7.3 PKN fracture geometry. Fracture height, length, and largest aperture are indicated in the figure (L, H, and w_{max}, respectively).

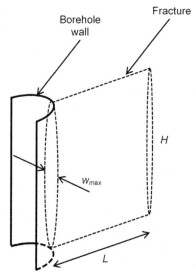

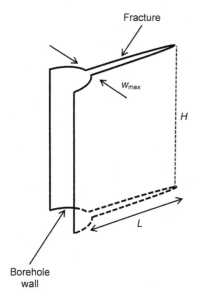

Fracture

w_{max}

H

L

Borehole
wall

FIGURE 7.4 KGD fracture geometry. Fracture
height, length, and largest aperture are indicated in
the figure (L, H, and w_{max}, respectively).

(or KGD) as a model of natural fractures implies, however, that the borehole is
drilled precisely along the fracture plane. In reality, the borehole can meet a
natural fracture at any angle.

In the KGD model, the fracture aperture at the borehole wall is constant
along the well and is given by [31].

$$w_{max} = \frac{4L(P_w - \sigma_h)}{E'} \qquad [7.4]$$

where L is the fracture length.

Evaluating the fracture aperture by means of either the PKN or the KGD
model requires that the elastic properties of the rock are known. It also requires
that either L or H is known. The fracture height could be estimated by means
of logging. The fracture length may then be estimated from the volume of fluid
lost into the fracture. However, even when these data are available, there are
major uncertainties pertaining to the use of these two models. For instance,
losses may be caused by multiple induced fractures; the rock behavior,
assumed elastic when deriving Eqs. [7.2], [7.3], and [7.4], is rarely such.
Moreover, in general, neither PKN nor KGD represents realistic fracture geometry; eg, fractures induced from deviated wells are often nonplanar.

The limited capacity of PKN and KGD models and the lack of viable
alternatives are the reasons why the fracture aperture is a major uncertainty in
lost circulation treatment design even when lost circulation is due to induced
fractures.

7.13 CURING LOSSES OF DIFFERENT SEVERITY

Information about the loss mechanism is often unavailable when the losses occur. The treatment must then be based solely on the severity of losses; ie, the number of barrels lost per hour. If both types of information—ie, the mechanism and the flow rate—are available, both should be analyzed and used when choosing and designing the treatment.

When seepage losses occur, they can be treated with LCM, or the decision can be made to drill ahead with losses. Treating losses might be necessary if they otherwise lead to other drilling problems, such as stuck pipe.

Similarly, when partial losses occur, they can be treated, or the decision can be made to drill ahead with losses. Treating losses is necessary if they otherwise lead to other drilling problems, or if the cost of the drilling fluid lost into the formation becomes significant [5]. If pumping LCM does not stop the losses, settable treatments can be used (cement, gunk, cross-linked pills, etc.).

Severe losses may occur in different types of formations and can be caused by natural fractures, induced fractures, vugs, or high-permeability matrix. When severe losses occur, they should be cured because severe losses may jeopardize well control. LCM, typically pumped as a pill, can be used to treat severe losses. If LCM does not help, settable systems can be used (cement, gunk, cross-linked pills, etc.) [9], or the decision can be made to use mud-cap drilling and drill ahead with no returns to surface [13].

Total losses (no returns to surface) are the most troublesome type of lost circulation. They may occur in different types of formations, particularly in vugular rocks, naturally fractured carbonates, and in geothermal drilling. Unconsolidated high-permeability formations may also cause total losses. Total losses must be treated in order to maintain well control. An attempt can be made to use LCM (typically pumped as a pill), but LCM alone rarely can cure total losses. If LCM does not help, a settable system can be tried. If all these treatments fail to stop losses, mud-cap drilling can be implemented to drill without returns [13].

In any of the above scenarios, reducing the ECD should be considered the most effective first option that could immediately cure or mitigate the problem. This can be achieved, for example, by reducing the static mud weight or the pump rate. Reducing the ECD should, however, be implemented carefully so as to avoid influx of formation fluids when circulation is stopped. Moreover, reducing the pump rate should not impair the hole cleaning.

7.14 LABORATORY TESTING OF LOST CIRCULATION MATERIALS AND LOSS PREVENTION MATERIALS

As in other branches of engineering, laboratory testing of proposed technologies is crucial for making progress in the realm of lost circulation treatments, too. A variety of methods have been proposed for evaluating LCMs and LPMs.

The quality of LCMs is usually assessed based on two properties [2]:

- total 30-min fluid loss (the lesser the better);
- time required to make a seal (the quicker the better).

These parameters can be evaluated by using filtration into a porous media (a ceramic disk or a rock sample) or into a slot representing a fracture.

The quality of LPMs is additionally characterized by [2]:

- increase of the formation breakdown pressure, fracture propagation pressure, and/or fracture reopening pressure;
- the maximum differential pressure that the seal can sustain (the higher the better);
- particle strength and resiliency (the greater the better).

As a result of lacking consensus about the mechanism of wellbore strengthening, different properties of LPMs may be considered more important by different researchers.

Quantifying the bridging and sealing capacity of LCMs is crucial for optimizing the treatments. The heart of the devices used for testing particle bridging and sealing capacity is a porous or slotted disk [12]. Examples of slotted disks are schematically shown in Fig. 7.5.

A particle plugging apparatus described in refs. [10,15] enables quantification of fracture bridging and sealing capacity of particulate and fiber LCMs and LPMs. The fluid with suspended particles is injected into a slot representing a fracture of a variable aperture. The slot aperture is tapered from 2500 to 1000 µm. As pointed out in Ref. [15], a tapered slot is more representative of an induced fracture than a slot of a constant aperture would be. The tapered-slot apparatus is suitable, for example, for testing wellbore strengthening materials (WSMs) intended for bridging and sealing of induced fractures.

The pressure gradient is applied in the tapered-slot apparatus so as to make the fluid flow upwards. This eliminates the possible error caused by particle settling that would otherwise overestimate the bridging and sealing capacity of the material.

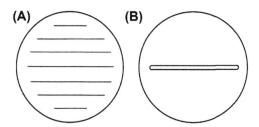

FIGURE 7.5 Schematic plots of slotted disks used to evaluate the LCM's/LPM's capacity to bridge and seal fractures. Disks with multiple slots (A) and a single slot (B) can be used.

A device with a tapered slot was used in laboratory tests of WSMs in Ref. [32]. The slot aperture was equal to 1 mm at the wide end. At the end representing the fracture tip, the slot aperture was equal to 0 (model of a closed fracture) or 0.5 mm (model of an opened fracture). The other dimensions of the slot were as follows: fracture length from mouth to tip was equal to 178 mm; fracture height was equal to 38 mm. The fracture faces were made of sandstone. Temperature control was enabled. The apparatus could thereby approximate a real fracture quite well.

As discussed in Chapter "Natural Fractures in Rocks," roughness of real fractures affects particle transport and deposition. A particle plugging apparatus would therefore be more representative of a real fracture if the surfaces of the slot were rough. Roughness of real fractures facilitates bridging and sealing. Therefore, laboratory devices with smooth-walled slots underestimate the bridging and sealing capacity of LCMs and LPMs.

A device for laboratory testing of flowing and sealing properties of LCM described in Ref. [4] consists of a metal tube with a piston at its inflow end and a 5-mm slot at the outflow end. The slot simulates a fracture. The tube is filled with the material (drilling fluid with LCM) to be tested, and a constant flow rate is applied displacing the fluid through the slot. Pressure difference is monitored as a function of the volume pumped. When LCM fails to bridge the fracture, all fluid and solids flow out through the slot, and the pressure does not build up (Fig. 7.6, dotted line). If solids bridge the slot, but fail to seal it, the fluid keeps flowing out, and the pressure gradient increases gradually as solids are accumulating in the slot (Fig. 7.6, dashed line). If solids create a low-permeability seal across the slot, the pressure gradient

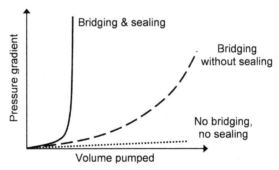

FIGURE 7.6 Schematic plot of pressure versus pumped fluid volume in a laboratory test of flowing and sealing capacity of LCM. Three cases are shown: no bridging and no sealing (dotted line); bridging but no sealing (dashed line); bridging and sealing (solid line). *Based on Droger N, Eliseeva K, Todd L, Ellis C, Salih O, Silko N, et al. Degradable fiber pill for lost circulation in fractured reservoir sections. IADC/SPE paper 168024 presented at the 2014 IADC/SPE Drilling Conference and Exhibition held in Fort Worth, Texas, USA, March 4–6, 2014.*

increases rapidly, with no solids or fluid eventually exiting the slot (Fig. 7.6, solid line). The LCM is considered to be effective in this case.

A modified fluid-loss cell was proposed in Ref. [4] to study the behavior of a fluid with LCM at higher flow rates as it flows through the bottomhole assembly. In this device, the fluid is channeled through a slot (1−5 mm wide) or a nozzle (7/32 to 16/32 in) under controlled applied pressure gradient. Monitoring the flow rate, volume, and constituency of the fluid coming out of the cell provides an indication of bridging the annular restrictions and nozzles.

In the permeability plugging apparatus used by Scorsone et al. [3], a particulate material was deposited on a high-permeability aloxite disk. The differential pressure that the so-obtained seal could sustain was found to be equal to 1500 psi (10 MPa). It was concluded that squeezing this material into the lost circulation zone could increase the fracturing pressure by a similar amount.

In Ref. [33], bridging and sealing of fractures in porous media was studied by means of an apparatus in which an LCM-carrying drilling fluid was injected into the gap between two porous plates (Fig. 7.7). The plates were made of a synthetic porous material (permeability c. 100 mD). The gap between the plates represented a fracture. A blend of graphite and nutshells (or graphite and calcium carbonate) was used as LCM. The outer edge of the "fracture" could be either closed or opened for flow out. The gap width was set equal to 200, 500, or 1000 μm and was held constant in each experiment [33]. Such setup could be representative of, for example, a natural fracture. A short induced fracture, on the other hand, might be better modeled with a tapered-slot apparatus discussed earlier in this section. During the test, the fluid entering through a hole in one of the plates could flow through the fracture if the fracture edge was opened. Part of the fluid could leak through the fracture wall into the porous media. Independent control of these two flows was enabled. An

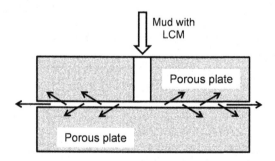

FIGURE 7.7 Principle of testing the ability of LCM to bridge and seal fractures. *Based on Kaageson-Loe N, Sanders MW, Growcock F, Taugbøl K, Horsrud P, Singelstad AV, et al. Particulate-based loss-prevention material—the secrets of fracture sealing revealed! SPE Drill Completion 2009;24(4):581−9.*

increase in the injection pressure was used as an indicator of a seal established inside the gap. This setup allowed Kaageson-Loe et al. [33] to perform a thorough study of bridging and sealing, and investigate different factors that affect these phenomena. In particular, the maximum size of LCM particles was found to impact the sealing ability profoundly.

Another device of this type was described by Sanders et al. [34]. The gap between two matching corrugated aluminum platens represented a fracture. The diameter of the platens was 6.35 cm. The platens were sandblasted in order to create some roughness. Since the platens had zero porosity and permeability, the device was particularly suitable for studying fracture sealing in low-permeability rocks; eg, shale or tight sand. Sanders et al. pointed out that using aluminum instead of a real low-permeability rock improved the reproducibility of the test results. During the experiment, drilling fluid containing LPM or LCM was injected between the plates. A constant fluid pressure of 25 psi (0.17 MPa) was applied at the "fracture tip"; ie, the outer edge of the gap between the platens. The geometry was similar to the one shown in Fig. 7.7. A confining pressure was applied on the platens to create the "fracture closure pressure" on the order of 125 psi (0.86 MPa). The following parameters were monitored during the experiment: the fluid loss through the "fracture tip," the injection pressure, and the variation of the gap aperture (opening/closing of the "fracture"). The aperture was 280 μm−1.1 mm in the experiments, typically around 500 μm. An increase in the injection pressure and a decrease in the fluid loss rate signified that a seal was formed inside the gap.

The devices of the type described in refs. [33,34] and schematically illustrated in Fig. 7.7 enable discrimination between effective and ineffective particulate materials. Materials with higher pressure buildup and lower fluid loss are, in general, more effective.

Larger apparatuses for testing fracture sealing have also been developed [34,35]. For instance, a high-pressure fracture test rig described in [34] operates on a 15-cm rock core. The maximum injection pressure can be up to 11,000 psi (75.8 MPa). Fracture initiation, propagation, and reopening can be studied in this rig.

A transparent cell for visually monitoring transport of LCM/LPM particles in a slot representing a fracture was described in [35].

Introduction of degradable LCMs necessitated the development of laboratory techniques that would enable users and manufacturers to quantify stability of such materials at downhole conditions, in particular temperature. To this end, an experiment was proposed where degradable LCM is placed into a metal tube, the tube is placed in an oven, and fluid flow is established through the tube [4]. A sudden drop in the pressure gradient indicates breakdown of the material. The time to breakdown—ie, the time during which the LCM can maintain its sealing capacity—characterizes the LCM stability at the given temperature [4].

Laboratory testing of shear-induced degradation of LCM particles involves blending and shearing of particles in a fluid using a mixer. The change in size, shape, and mass of LCM particles caused by abrasion during this mixing process can be used to quantify the particle degradation under shear [16].

Several types of laboratory tests have been used to characterize the mechanical strength of a seal. Mechanical strength is crucial for the seal's ability to sustain differential pressure and mechanical loads without breaking apart and without increasing permeability by crack development. Two parameters commonly used to characterize the mechanical stability of sealing materials are the shear strength and the UCS.

A push-out test has been used to quantify the shear strength. A seal of a specified thickness, h, can be built, for example, by filtration on a porous disc [3,11]. The defluidization time of the material can be evaluated during the seal building. The seal is then placed into the push-out testing device, and the push-out test is performed by moving the loading piston down, inducing shear on a cylindrical surface of diameter d (Fig. 7.8). The test is similar to the shale puncher used in rock mechanics to evaluate the shear strength of shales using small disc-shaped shale specimens [36]. The test is also similar to the push-out test used to characterize bonding strength at cement–steel and cement–rock interfaces [5]. The shear strength of the seal, σ_s, is evaluated from the peak load measured during the test, F_{max}, as follows:

$$\sigma_s = \frac{F_{max}}{\pi h d} \qquad [7.5]$$

Another test commonly used to quantify the mechanical strength of the seal is the uniaxial compressive test. A cylindrical sample having the

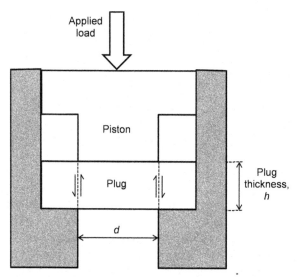

FIGURE 7.8 Principle of push-out test of an LCM seal. The shear strength of the plug is calculated from Eq. [7.5]. *Based on Sanders MW, Scorsone JT, Friedheim JE. High-fluid-loss, high-strength lost circulation treatments. SPE paper 135472 presented at the SPE Deepwater Drilling and Completions Conference held in Galveston, Texas, USA, October 5–6, 2010.*

height-to-diameter ratio equal to two is subjected to an increasing axial load until it fails. The specimen typically fails by developing multiple fractures approximately parallel to or inclined to the loading axis. The maximum compressive stress reached in the test yields the UCS of the material. The UCS of an LCM seal can be on the order of 70 psi (0.5 MPa) [11]. Some materials—eg, fiber-laden LCMs—exhibit a substantial compressibility without a well-pronounced peak load in the UCS test [11].

Mechanical strength, represented by UCS, is an important property of particulate materials. Performance of settable materials depends, in addition, on their adhesive properties. For instance, settable materials used in wellbore strengthening in shales must adhere to the shale surface so that the seal is not easily displaced back into the wellbore when the wellbore pressure is decreased. A special laboratory test has been designed in order to evaluate this property [37]. Development of new lost circulation solutions calls for new, nonstandard experimental techniques to verify and optimize the performance of these solutions in controlled laboratory conditions before taking them into the field.

Another parameter that describes mechanical properties of particulates is *resiliency* introduced in Chapter "Preventing Lost Circulation." A resilient material is capable of restoring its size after compressive load has been removed. A compressive stress of 10,000 psi (69 MPa) is applied to an LCM sample. The load is then removed, and the percentage gain in the sample's height is measured. This gain is a measure of the material's resiliency. If H_1 is the height of the specimen under 10,000 psi (69 MPa), and H_2 is its height after the load has been removed, the resiliency is defined as $100\% \cdot (H_2-H_1)/H_1$ [15].

Crushing resistance is another property that describes the mechanical behavior of an LPM bridge or seal. It can be measured in the following laboratory test described in Ref. [15]: A sample of LPM is analyzed to obtain its initial PSD and, in particular, the values of D_{50} and D_{90}. The sample is then subjected to a compressive load. Upon unloading, the sample is analyzed again, and its PSD is measured. The percentage change of D_{50} and D_{90} compared with their original values provides a measure for the crushing resistance of the material. It was shown in Ref. [15] that crushing resistance of a particulate material could be improved by mixing it with a material that has better crushing resistance. For instance, the crushing resistance of ground marble can be improved by mixing it with resilient graphitic carbon.

In addition to laboratory testing, yard tests are performed on LCM materials to estimate, for example, their ability to pass through flow restrictions, such as bit nozzles or constrictions in the bottomhole assembly [3,11]. In particular, in the jet nozzle test, LCM is pumped through a nozzle in order to confirm that it will not plug the bit nozzles during the in situ placement. Flow rates of 100, 200, and 400 gal/min were employed in the jet nozzle tests in Ref. [3]. Similar yard tests can be performed to make sure that LCM will not plug the bottomhole assembly.

7.15 SUMMARY

Three kinds of treatments are used to cure lost circulation: LCMs (particulates and fibers), settable materials, and blends of the two. Settable materials often need some extra preparation time and also some setting time downhole before they can stop the losses. These materials may work in formations where LCMs are ineffective. LCMs can be pumped with the drilling fluid and can start curing losses immediately once they have reached the thief zone.

When choosing and designing the treatment, all available information should be analyzed and used. This includes the loss severity (barrels per hour), the loss mechanism (ie, induced or natural fractures, vugs, or high-permeability rock), quantitative data about fractures and pores (fracture apertures, spacing, pore throat size, etc.), geological setting (pay zone, caprock, shale, unconsolidated sand, gravel, etc.).

Usually, reducing the ECD should be tried first. If this does not help, other actions can be taken. Some examples of such actions are provided in Tables 7.1 and 7.2. Usually, LCMs are tried as the first cure. Calcium carbonate, nut shells, graphite, and fibers are among the most common LCMs currently used in oil and gas industry. More severe losses require coarser particulates and higher solids volume fractions. Properly designed blends of different LCMs (eg, calcium carbonate + fibers) are often more effective than single LCM (eg, calcium carbonate alone). If LCM fails to cure the losses, settable materials can be spotted (cross-linked pills, gunk, cement, etc.). If that does not help, the decision can be made to drill ahead with losses, or to implement mud-cap drilling.

TABLE 7.1 Examples of Actions Based on Loss Severity

Severity of Lost Circulation	Possible Actions and Treatments
Seepage (<10 bbl/h)	• Do nothing, drill ahead (solids present in the mud will hopefully create a seal) • Stop drilling, wait for the hole to "heal" (a few hours) • Treat the entire system with fine particulate/fiber LCM (calcium carbonate, graphite, gilsonite, fibers, etc.; 5–15 lb/bbl in total; ie, c. 14.25–42.75 kg/m³, in total) • High-fluid-loss squeeze
Partial (10–100 bbl/h)	• Treat the entire system with particulate/fiber LCM • Circulate a particulate/fiber LCM pill • High-fluid-loss squeeze • Two-fluid settable systems • Cross-linked systems • Cement • Gunk

Continued

TABLE 7.1 Examples of Actions Based on Loss Severity—cont'd

Severity of Lost Circulation	Possible Actions and Treatments
Severe (>100 bbl/h)	• Particulate/fiber LCM • High-fluid-loss squeeze • Cross-linked systems • Two-fluid settable systems • Cement • Gunk • Drilling blind • Mud-cap drilling
Total (no returns)	• Drilling blind • Mud-cap drilling • Cross-linked systems • Cement • Gunk

TABLE 7.2 Examples of Actions Based on Loss Mechanisms

Mechanism of Lost Circulation	Possible Actions and Treatments
Losses into porous matrix	• Drill ahead (solids present in mud will hopefully create a seal) • Particulate/fiber LCM • High-fluid-loss squeeze
Losses in vugular/cavernous formations	• Drill ahead, if possible • Cross-linked systems • Cement
Losses into natural fractures	• Particulate/fiber LCM • High-fluid-loss squeeze • Cross-linked systems • Two-fluid settable systems • Cement • Gunk • Drilling blind • Mud-cap drilling
Losses into induced fractures	• Particulate/fiber LCM • Cross-linked systems

REFERENCES

[1] Messenger JU. Lost circulation. PennWell Books; 1981. p. 112.

[2] Alsaba M, Nygaard R, Hareland G, Contreras O. Review of lost circulation materials and treatments with an updated classification. AADE-14-FTCE-25 paper presented at the 2014 AADE fluids Technical Conference and Exhibition held at the Hilton Houston North Hotel, Houston, Texas; April 15−16, 2014.

[3] Scorsone JT, Dakin ES, Sanders MW. Maximize drilling time by minimizing circulation losses. AADE-10-DF-HO-42 presented at the 2010 AADE fluids Conference and Exhibition held at the Hilton Houston North, Houston, Texas; April 6−7, 2010.

[4] Droger N, Eliseeva K, Todd L, Ellis C, Salih O, Silko N, et al. Degradable fiber pill for lost circulation in fractured reservoir sections. IADC/SPE paper 168024 presented at the 2014 IADC/SPE Drilling Conference and Exhibition held in Fort Worth, Texas, USA; March 4−6, 2014.

[5] Nelson EB, Guillot D. Well cementing. 2nd ed. Schlumberger; 2006. p. 773.

[6] Ivan C, Bruton JR, Bloys B. How can we best manage lost circulation? AADE-03-NTCE-38 paper presented at the AADE 2003 National Technology Conference "Practical solutions for drilling challenges," held at the Radisson Astrodome Houston, Texas; April 1−3, 2003 [in Houston, Texas].

[7] Goncalves R, Tajalie AF, Meyer A. Overcoming lost circulation while cementing riserless tophole in deepwater. OTC-24764-MS paper presented at the Offshore Technology Conference Asia held in Kuala Lumpur, Malaysia; March 25−28, 2014.

[8] Whitfill DL, Hemphill T. All lost-circulation materials and systems are not created equal. SPE paper 84319 presented at the SPE Annual Technical Conference and Exhibition held in Denver, Colorado, USA; October 5−8, 2003.

[9] Alsaba M, Nygaard R, Saasen A, Nes O-M. Lost circulation materials capability of sealing wide fractures. SPE paper 170285 presented at the SPE Deepwater Drilling and Completions Conference held in Galveston, Texas, USA; September 10−11, 2014.

[10] Kumar A, Savari S, Jamison DE, Whitfill DL. Application of fiber laden pill for controlling lost circulation in natural fractures. AADE-11-NTCE-19 paper presented at the 2011 AADE National Technical Conference and Exhibition held at Hilton Houston North Hotel, Houston, Texas; April 12−14, 2011.

[11] Sanders MW, Scorsone JT, Friedheim JE. High-fluid-loss, high-strength lost circulation treatments. SPE paper 135472 presented at the SPE Deepwater Drilling and Completions Conference held in Galveston, Texas, USA; October 5−6, 2010.

[12] Zeilinger S, Dupriest F, Turton R, Butler H, Wang H. Utilizing an engineered particle drilling fluid to overcome coal drilling challenges. IADC/SPE paper 128712 presented at the 2010 IADC/SPE Drilling Conference and Exhibition held in New Orleans, Louisiana, USA; February 2−4, 2010.

[13] Savari S, Whitfill DL. Managing losses in naturally fractured formations: sometimes nano is too small. SPE/IADC paper 173062 presented at the SPE/IADC Drilling Conference and Exhibition held in London, United Kingdom; March 17−19, 2015.

[14] Onyia EC. Experimental data analysis of lost-circulation problems during drilling with oil-based muds. SPE Drill Completion 1994;9(1):25−31.

[15] Kumar A, Savari S. Lost circulation control and wellbore strengthening: looking beyond particle size distribution. AADE-11-NTCE-21 paper presented at the 2011 AADE National Technical Conference and Exhibition held at the Hilton Houston North Hotel, Houston, Texas; April 12−14, 2011.

[16] Scott PD, Beardmore DH, Wade ZD, Evans E, Franks KD. Size degradation of granular lost circulation materials. IADC/SPE paper 151227 presented at the 2012 IADC/SPE Drilling Conference and Exhibition held in San Diego, California, USA; March 6–8, 2012.

[17] Adachi J, Bailey L, Houwen OH, Meeten GH, Way PW, Schlemmer RP. Depleted zone drilling: reducing mud losses into fractures. IADC/SPE paper 87224 presented at the IADC/SPE Drilling Conference held in Dallas, Texas, USA; March 2–4, 2004.

[18] Drilling Specialties Company. Lost circulation guide. 2014.

[19] Lécolier E, Herzhaft B, Néau L, Quillien B, Kieffer J. Development of a nanocomposite gel for lost circulation treatment. SPE paper 94686 presented at the SPE European formation damage Conference held in Scheveningen, The Netherlands; May 25–27, 2005.

[20] Wang H, Sweatman R, Engelman B, Deeg W, Whitfill D, Soliman M, et al. Best practice in understanding and managing lost circulation challenges. SPE Drill Completion 2008;23(2):168–75.

[21] Fidan E, Babadagli T, Kuru E. Use of cement as lost circulation material—field case studies. IADC/SPE paper 88005 presented at the IADC/SPE Asia Pacific drilling Technology Conference and Exhibition held in Kuala Lumpur, Malaysia; September 13–15, 2004.

[22] Al Maskary S, Abdul Halim A, Al Menhali S. Curing losses while drilling & cementing. SPE paper 171910 presented at the Abu Dhabi International petroleum Exhibition and Conference held in Abu Dhabi, UAE; November 10–13, 2014.

[23] Mata F, Veiga M. Crosslinked cements solve lost circulation problems. SPE paper 90496 presented at the SPE Annual Technical Conference and Exhibition held in Houston, Texas, USA; September 26–29, 2004.

[24] Caughron DE, Renfrow DK, Bruton JR, Ivan CD, Broussard PN, Bratton TR, et al. Unique crosslinking pill in tandem with fracture prediction model cures circulation losses in deepwater Gulf of Mexico. IADC/SPE paper 74518 presented at the IADC/SPE Drilling Conference held in Dallas, Texas; February 26–28, 2002.

[25] Ferras M, Galal M, Power D. Lost circulation solutions for severe sub-salt thief zones. Paper AADE-02-DFWM-HO-30 presented at the AADE 2002 Technology Conference "Drilling & completion fluids and Waste Management," held at the Radisson Astrodome, Houston, Texas; April 2–3, 2002 [in Houston, Texas].

[26] Savari S, Whitfill DL, Scorsone JT. Next-generation, right-angle-setting composition for eliminating total lost circulation. SPE/IADC paper 166697 presented at the SPE/IADC Middle East Drilling Technology Conference and Exhibition held in Dubai, UAE; October 7–9, 2013.

[27] Power D, Ivan CD, Brooks SW. The top 10 lost circulation concerns in deepwater drilling. SPE paper 81133 presented at the SPE Latin American and Caribbean petroleum engineering Conference held in Port-of-Spain, Trinidad, West Indies; April 27–30, 2003.

[28] Ghalambor A, Salehi S, Shahri MP, Karimi M. Integrated workflow for lost circulation prediction. SPE paper 168123 presented at the SPE International Symposium and Exhibition on Formation Damage Control held in Lafayette, Louisiana, USA; February 26–28, 2014.

[29] Liétard O, Unwin T, Guillot DJ, Hodder MH. Fracture width logging while drilling and drilling mud/loss-circulation-material selection guidelines in naturally fractured reservoirs. SPE Drill Completion 1999;14(3):168–77.

[30] Dyke CG, Wu B, Milton-Tayler D. Advances in characterizing natural-fracture permeability from mud-log data. SPE Form Eval 1995;10(3):160–6.

[31] Valkó P, Economides MJ. Hydraulic fracture mechanics. Chichester: John Wiley & Sons; 1995. p. 298.

[32] Aston MS, Alberty MW, McLean MR, de Jong HJ, Armagost K. Drilling fluids for wellbore strengthening. IADC/SPE paper 87130 presented at the IADC/SPE Drilling Conference held in Dallas, Texas, USA; March 2−4, 2004.

[33] Kaageson-Loe N, Sanders MW, Growcock F, Taugbøl K, Horsrud P, Singelstad AV, et al. Particulate-based loss-prevention material—the secrets of fracture sealing revealed! SPE Drill Completion 2009;24(4):581−9.

[34] Sanders MW, Young S, Friedheim JE. Development and testing of novel additives for improved wellbore stability and reduced losses. AADE-08-DF-HO-19 paper presented at the 2008 AADE Fluids Conference and Exhibition held at the Wyndam Greenspoint Hotel, Houston, Texas; April 8−9, 2008.

[35] Guo Q, Cook J, Way P, Ji L, Friedheim JE. A comprehensive experimental study on wellbore strengthening. IADC/SPE paper 167957 presented at the 2014 IADC/SPE Drilling Conference and Exhibition held in Fort Worth, Texas, USA; March 4−6, 2014.

[36] Stenebråten JF, Fjær E, Haaland S, Lavrov AV, Sønstebø EF. The shale puncher—a compact tool for fast testing of small shale samples. 42nd US rock mechanics Symposium and 2nd US Canada rock mechanics Symposium held in San Francisco; 29 June−2 July, 2008.

[37] Aston MS, Alberty MW, Duncum S, Bruton JR, Friedheim JE, Sanders MW. A new treatment for wellbore strengthening in shale. SPE paper 110713 presented at the 2007 SPE Annual Technical Conference and Exhibition held in Anaheim, California, USA; November 11−14, 2007.

Chapter 8

Knowledge Gaps and Outstanding Issues

Significant progress has been made in preventing and combatting lost circulation over the past three decades. Still, several knowledge gaps and outstanding challenges persist.

Probably the best way to deal with lost circulation is to not let it happen in the first place. This can be achieved by keeping the bottomhole pressure below the lost-circulation pressure. However, predicting the lost-circulation pressure is not an easy task. We saw in chapter "Mechanisms and Diagnostics of Lost Circulation" that lost-circulation pressure may be equal to the in situ pore pressure, the minimum principal in situ stress, or a function of all three in situ principal stresses, depending on the lost-circulation mechanism. In addition, the lost-circulation pressure may depend on the apertures and orientations of natural fractures. Obtaining these data is difficult (if not impossible) until the formation has been exposed by drilling the well. Even then, though, the minimum in situ stress is usually measured only at some locations in the well, typically at the casing shoe. Other data, for instance, fracture apertures, are not measured at all. Improving formation characterization, in particular stress measurements and evaluation of fracture properties (aperture, connectivity, etc.), will improve the performance of lost-circulation treatments and will make their outcome more consistent. Better predrill characterization of the formation will improve the effect of loss-prevention treatments.

One such preventive treatment is wellbore strengthening (chapter: Preventing Lost Circulation). Designing strengthening treatments requires that a reliable predictive theory is available for this technology. Different theories often end up with opposite recommendations regarding the choice of the particle size distribution and solids concentration. Developing a commonly accepted theory of wellbore strengthening that could be used in different types of rocks and with different muds is an outstanding task. As pointed out in Ref. [1], the theory of wellbore strengthening should incorporate the fluid dynamics in the fracture, the deformation of the surrounding rock, the leakoff through the fracture wall, and the particle transport and deposition in the fracture. This will make the model quite complex and computationally demanding. Simpler, engineering-type models should be developed as well once the wellbore strengthening mechanisms are fully understood. Laboratory

Lost Circulation. http://dx.doi.org/10.1016/B978-0-12-803916-8.00008-X

tests are indispensable for testing theoretical models and providing the input data. Understanding the mechanisms of wellbore strengthening is necessary in order to establish generally accepted laboratory tests of loss-prevention materials (LPMs) [2].

Application of wellbore strengthening has been limited to high-permeability rocks so far. In low-permeability rocks, the method performs poorly. In shales, wellbore strengthening usually does not work with particulate LPMs and requires the use of settable materials (eg, cement or crosslinked polymers). Further development of loss-prevention techniques for low-permeability rocks is an outstanding task. Low-permeability rocks such as shale, and vugular and naturally fractured rocks such as carbonates, are often singled out as the greatest challenges with respect to lost-circulation prevention [3].

This brings about also the question of what the "low-permeability rocks" really are. It is currently not clear what is the minimum formation permeability for which LPMs still can work [4], and how lost-circulation materials (LCMs) could be optimized to be more effective in low-permeability rocks.

The volume of LPM or settable material that needs to be pumped in a squeeze is another uncertainty in wellbore strengthening. This uncertainty is due to unknown geology, number, and volume of natural fractures, unknown in situ stresses, etc. This results in preventive treatments having only partial success. Additional treatments may then be required to ensure the desired strengthening effect [5]. Better characterization of thief zones and better treatment design based on the available information are outstanding issues in loss prevention.

The prevention of lost circulation can be improved by better well planning. In particular, rocks around faults may represent potential thief zones. This issue is not that simple, though, since some faults give rise to drilling problems, while others do not [6]. Better discrimination between potentially troublesome and potentially harmless geological structures will make wells more cost-effective.

Once losses have occurred, interpretation and differential diagnosis of their mechanisms are crucial for treatment design. We saw in chapter "Mechanisms and Diagnostics of Lost Circulation" that differential diagnosis of losses can be improved by using high-frequency flow rate measurements. This requires installation of flowmeters, which is not routinely done. Moreover, even if such flowmeters are installed and the high-frequency flow rate data become available, there is ambiguity in the interpretation of the "flow rate versus time" curves. There have been lost-circulation incidents where such interpretation was straightforward, especially if other drilling data were available as well (eg, the standpipe pressure, the rate of penetration, the weight on bit, etc.). However, even when flow rates and other drilling data are available, the mechanism of losses (porous matrix, vugular zones, natural fractures, or induced fractures) sometimes cannot be reliably identified [7]. The lack of generally applicable measurement techniques and reliable interpretation procedures introduce uncertainty into the design of lost-circulation prevention and treatments.

Curing severe and total losses is still a major challenge in the industry [8]. No cure-all solution to this challenge has been found yet. Even though effective treatments do exist for such losses (eg, cement squeeze, gunk squeeze, polymer pills), they usually cannot be applied in reservoir sections because of formation damage they cause. Therefore, mud losses still are a major problem in heavily fractured reservoirs with narrow drilling margins, such as naturally fractured deepwater or depleted reservoirs. Sealing fractures that are wider than 5 mm is another outstanding problem in LPM and LCM design [9].

Despite the challenges outlined above, continuous progress in lost-circulation research and in developing more effective treatments makes it possible to drill deeper and more difficult wells every year. Therefore, solutions to today's challenges will most likely be found in the years to come.

REFERENCES

[1] van Oort E, Razavi SO. Wellbore strengthening and casing smear: the common underlying mechanism. IADC/SPE paper 168041 presented at the 2014 IADC/SPE Drilling Conference and Exhibition held in Fort Worth, Texas, USA, 4−6 March 2014.

[2] Alsaba M, Nygaard R, Hareland G, Contreras O. Review of lost circulation materials and treatments with an updated classification. AADE-14-FTCE-25 paper presented at the 2014 AADE Fluids Technical Conference and Exhibition held at the Hilton Houston North Hotel, Houston, Texas, April 15−16, 2014.

[3] Kumar A, Savari S. Lost circulation control and wellbore strengthening: looking beyond particle size distribution. AADE-11-NTCE-21 paper presented at the 2011 AADE National Technical Conference and Exhibition held at the Hilton Houston North Hotel, Houston, Texas, April 12−14, 2011.

[4] Dupriest FE. Fracture closure stress (FCS) and lost returns practices. SPE/IADC paper 92192 presented at the SPE/IADC Drilling Conference held in Amsterdam, The Netherlands, 23−25 February 2005.

[5] Traugott D, Sweatman R, Vincent R. Increasing the wellbore pressure containment in Gulf of Mexico HP/HT wells. SPE Drill Completion 2007;22(1):16−25.

[6] Kristiansen TG, Mandziuch K, Heavey P, Kol H. Minimizing drilling risk in extended-reach wells at Valhall using geomechanics, geoscience and 3D visualization technology. SPE/IADC paper 52863 presented at the 1999 SPE/IADC Drilling Conference held in Amsterdam, Holland, 9−11 March 1999.

[7] Beda G, Carugo C. Use of mud microloss analysis while drilling to improve the formation evaluation in fractured reservoir. SPE paper 71737 presented at the 2001 SPE Annual Technical Conference and Exhibition held in New Orleans, Louisiana, 30 September−3 October 2001.

[8]   Savari S, Whitfill DL. Managing losses in naturally fractured formations: sometimes nano is too small. SPE/IADC paper 173062 presented at the SPE/IADC Drilling Conference and Exhibition held in London, United Kingdom, 17−19 March 2015.

[9]   Droger N, Eliseeva K, Todd L, Ellis C, Salih O, Silko N, et al. Degradable fiber pill for lost circulation in fractured reservoir sections. IADC/SPE paper 168024 presented at the 2014 IADC/SPE Drilling Conference and Exhibition held in Fort Worth, Texas, USA, 4−6 March 2014.

Index

Printed in the United States
By Bookmasters